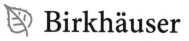

**Compact Textbooks in Mathematics**

This textbook series presents concise introductions to current topics in mathematics and mainly addresses advanced undergraduates and master students. The concept is to offer small books covering subject matter equivalent to 2- or 3-hour lectures or seminars which are also suitable for self-study. The books provide students and teachers with new perspectives and novel approaches. They may feature examples and exercises to illustrate key concepts and applications of the theoretical contents. The series also includes textbooks specifically speaking to the needs of students from other disciplines such as physics, computer science, engineering, life sciences, finance.

- **compact:** small books presenting the relevant knowledge
- **learning made easy:** examples and exercises illustrate the application of the contents
- **useful for lecturers:** each title can serve as basis and guideline for a semester course/lecture/seminar of 2-3 hours per week.

Mi-Ho Giga • Yoshikazu Giga

# A Basic Guide to Uniqueness Problems for Evolutionary Differential Equations

Mi-Ho Giga
Graduate School of Mathematical Sciences
The University of Tokyo
Tokyo, Japan

Yoshikazu Giga
Graduate School of Mathematical Sciences
The University of Tokyo
Tokyo, Japan

ISSN 2296-4568          ISSN 2296-455X   (electronic)
Compact Textbooks in Mathematics
ISBN 978-3-031-34795-5          ISBN 978-3-031-34796-2   (eBook)
https://doi.org/10.1007/978-3-031-34796-2

This book is published under the imprint Birkhäuser, www.birkhauser-science.com by the registered
company Springer Nature Switzerland AG
The registered company address is: Gewerbestrasse 11, 6330 Cham, Switzerland

Paper in this product is recyclable.

# Preface

It is of fundamental importance to study the uniqueness of a solution in the theory of partial differential equations, especially when the problem is nonlinear. Even if we focus on initial value problems for evolutionary differential equations, there is a large body of literature on fundamental equations.

This book is intended to serve as a short introduction to uniqueness questions for initial value problems, especially when one considers generalized solutions that may not be differentiable. For this purpose, we discuss three typical equations – ordinary differential equations, scalar conservation laws, and Hamilton–Jacobi equations.

Let us first consider the initial value problem of an autonomous system of ordinary differential equations of $x = x(t) \in \mathbf{R}^N$ of the form

$$\frac{dx}{dt} = b(x), \quad x(0) = X \in \mathbf{R}^N, \quad t \in I, \tag{1}$$

where $\mathbf{R}^N$ denotes the $N$-dimensional Euclidean space and $I$ denotes a time interval containing 0. Although a local-in-time solution exists when $b$ is continuous, it may not be unique. A conventional basic sufficient condition for uniqueness is that $b$ is Lipschitz continuous near $X$. However, there is a weaker and useful sufficient condition for the uniqueness called the Osgood condition. This book begins with such a uniqueness issue when $b$ is continuous.

Even if $b$ is discontinuous, it turns out that uniqueness for positive time holds when $b$ is monotone nonincreasing when the notion of a solution is defined in a suitable way. Actually, this forward uniqueness has wide applications to the theory of partial differential equations by considering the ordinary differential equation (1) in an infinite-dimensional Hilbert space rather than $\mathbf{R}^N$. This book only gives basic uniqueness issues for such problems.

If $b$ is neither monotone nonincreasing nor noncontinuous, in general, the solution may not be unique. However, if the derivative of $b$ has some integrability and the divergence of $b$ is zero, the solution operator

$$X \mapsto x(t, X)$$

can be uniquely determined to be almost all $X$. This type of theory was initiated by DiPerna and Lions [32] in the 1980s. It remains an active research area related to fluid dynamics. This book provides an introduction to the theory by focusing on the uniqueness part in a periodic setting. It also relates to the uniqueness issue of a linear transport equation,

$$\frac{\partial u}{\partial t} - b \cdot \nabla u = 0, \quad x \in \mathbf{R}^N, \quad t \in I,$$

when $b$ is not regular.

The uniqueness of a solution becomes more subtle for partial differential equations if a weak solution (a generalized solution that may not be differentiable) is considered. The first example is a scalar conservation law,

$$\frac{\partial u}{\partial t} + \frac{\partial}{\partial x} (f(u)) = 0, \quad x \in \mathbf{R}, \quad t \in I,$$

when $f$ is nonlinear. A typical example is the case where $f(u) = u^2/2$. In this case, the equation is called the Burgers equation. Even if the initial data are smooth, a solution may develop a jump discontinuity called a shock. A conventional notion of a weak solution may not guarantee uniqueness. An extra condition called the entropy condition guarantees its uniqueness. This book provides an introduction to the uniqueness of entropy solutions following the definition due to Kružkov [68]. A key idea of the uniqueness proof is doubling variables. Most of the discussion on conservation laws is taken from a book by Holden and Risebro [53].

The second example is a Hamilton–Jacobi equation:

$$\frac{\partial u}{\partial t} + H(x, \nabla u) = 0, \quad x \in \mathbf{R}^N, \quad t \in I,$$

where $H$ is a continuous function called a Hamiltonian. In a spatially one-dimensional case, if $H(x, p) = p^2/2$, then the Burgers equation is derived by differentiating the equation. Thus, it is expected that a smooth solution may be nondifferentiable in finite time. One should ask what the proper notion of a solution is. The notion of a viscosity solution introduced by Crandall and Lions [29] is a very suitable one to guarantee the uniqueness of a solution to an initial value problem for a Hamilton–Jacobi equation. A key idea is another version of "doubling variables." This book is intended to serve as a short introduction to the theory. If $H$ also depends on $u$ itself, it covers a scalar conservation law, so one must consider a discontinuous viscosity solution. The book concludes by studying viscosity solutions with a shock following [46].

The book was written for upper-level undergraduate students who plan on a pursuing graduate-level studies on one of the important aspects of differential equations. Of course, the book will help researchers in various fields to understand problems that mathematical analysts are typically concerned with. Although some basic knowledge of Sobolev spaces is necessary for Sect. 1.2 and Chap. 2, the book will be accessible to those unfamiliar with the details of the theory of function spaces. Basic terminology is listed in Appendix (Chap. 5). In principle, a basic mathematical term will appear for the first time in italics. If no explanation of the term appears in the nearby text, the reader is encouraged to consult the index to find the term on another page, where the term should be defined. For example, the word "measurable" first appears in Sect. 1.1.4, but its definition is given in Sects. 1.2.5 and 5.2. The reference is not exhaustive.

The book is organized as follows. Chapter 1 discusses the uniqueness problem for ordinary differential equations. Starting with the Lipschitz condition and the Oswood condition, Sect. 1.1 concludes by discussing equations with fractional time derivatives. Section 1.2 discussed the initial value problem for (1) where $-b$ is a maximal monotone operator in a Hilbert space. Chapter 2 discusses equation (1) for irregular divergence-free $b$ and the transport equation. Chapter 3 is devoted to the uniqueness of entropy solutions for scalar conservation laws. Chapter 4 presents an introduction to the theory of viscosity solutions for Hamilton–Jacobi equations. Chapter 5 presents basic terminology.

This book stems from course lectures by the second author given at the University of Tokyo for upper-level undergraduate students focusing on some uniqueness issues of solutions to differential equations. In nonlinear analysis, it often happens that it is more difficult to prove the uniqueness of a solution than the existence of a solution, especially when one considers a generalized solution. This is one reason why this topic was chosen.

In course lectures, part of the scalar conservation law is explained based on an excellent book by Holden and Risebro [53]. Thus, this part follows that book very closely, although a few simplifications are made.

The authors are grateful to Professor Nao Hamamuki, who gave us useful comments on the manuscript of this book. The authors thank Professor Hiroyoshi Mitake for providing some key references on Hamilton–Jacobi equations. The authors are also grateful to students who attended the course lectures at the University of Tokyo for their valuable comments. Last but not least, the authors would like to thank Ms. Satoko Kimura for preparing the final TeX file of this book.

The work of the second author was partly supported by the Japan Society for the Promotion of Science (JSPS), through grants KAKENHI, No. 20K20342, No. 19H00639, No. 18H05323, and No. 17H01091, and by Arithmer Inc. Daikin Industries, Ltd. and Ebara Corporation, through collaborative grants.

Tokyo, Japan                                                                                                                    Mi-Ho Giga
March 2023                                                                                                             Yoshikazu Giga

# Contents

# Uniqueness of Solutions to Initial Value Problems for Ordinary Differential Equations

In this chapter, we recall several types of sufficient conditions to guarantee the uniqueness of solutions to the initial value problem for ordinary differential equations. We first review classical Lipschitz- and Osgood-type uniqueness conditions. We then consider a gradient flow of a convex function that is not necessarily differentiable. This system may not satisfy the aforementioned uniqueness condition because the gradient of a convex function may not be continuous. Because of the monotonicity nature of the problem, we still find uniqueness.

We shall use standard notations. For $m = 1, 2, 3, \ldots$, let $\mathbf{R}^m$ denote the $m$-dimensional *Euclidean space*. In other words,

$$\mathbf{R}^m := \left\{ a = (a^1, \ldots, a^m) \mid a^i \in \mathbf{R} \text{ for } 1 \le i \le m \right\},$$

where $\mathbf{R}$ denotes the set of all real numbers. Here $a^i \in \mathbf{R}$ means that $a^i$ is an *element* of the set $\mathbf{R}$, i.e., $a^i$ is a real number in this case. By definition, $\mathbf{R}^1 = \mathbf{R}$. For two vectors $a = (a, \ldots, a^m) \in \mathbf{R}^m$ and $b = (b^1, \ldots, b^m) \in \mathbf{R}^m$, let $\langle a, b \rangle$ denote the standard *inner product* in $\mathbf{R}^m$ defined by

$$\langle a, b \rangle := \sum_{i=1}^m a^i b^i.$$

We also denote $\langle a, b \rangle$ simply by $a \cdot b$. Let $|a|$ denote the *Euclidean norm* defined by

$$|a| := \langle a, a \rangle^{1/2} = \sqrt{a \cdot a}.$$

If $m = 1$, then $|a|$ is simply the absolute value of $a$. For a vector $a \in \mathbf{R}^m$, the superscript $i$ of $a^i$ does not mean the power of $a$, unless $m = 1$. It represents the $i$th component.

© The Author(s), under exclusive license to Springer Nature Switzerland AG 2023
M.-H. Giga, Y. Giga, *A Basic Guide to Uniqueness Problems for Evolutionary
Differential Equations*, Compact Textbooks in Mathematics,
https://doi.org/10.1007/978-3-031-34796-2_1

## 1.1     Gronwall-Type Inequalities and Uniqueness of Solutions

We now consider an $N$-system of ordinary differential equations for $x = x(t)$ of the form

$$\dot{x} = b(x, t), \quad \dot{x} = dx/dt, \tag{1.1}$$

where $b = b(\cdot, t)$ is a time-dependent vector field in $\mathbf{R}^N$ and $t \in \mathbf{R}$. This is a concise form of the system

$$\frac{dx^i}{dt}(t) = b^i\left(x^1(t), \dots, x^N(t), t\right), \quad 1 \leq i \leq N,$$

where $x^i(t)$ $(1 \leq i \leq N)$ is an unknown function defined in some time interval containing the origin, and $b^i$ is a function of $N + 1$ variables. The concise form is obtained by setting vector-valued ($\mathbf{R}^N$-valued) functions $x(t) = \left(x^1(t), \dots, x^N(t)\right)$ and $b(x, t) = \left(b^1(x, t), \dots, b^N(x, t)\right)$.

Our main concern is whether or not a solution $x$ satisfying the initial condition, say, $x(0) = X$ for a given $X \in \mathbf{R}^N$, is unique.

### 1.1.1    Lipschitz Condition

A standard sufficient condition is the Lipschitz condition. For a given vector-valued ($\mathbf{R}^m$-valued) function $f = f(x) = \left(f^1(x), \dots, f^m(x)\right)$ defined in a subset $\Omega$ of $\mathbf{R}^N$, we set

$$[f]_{\mathrm{Lip}(\Omega)} = \sup\left\{|f(x) - f(y)|/|x - y| \ \middle| \ x, y \in \Omega, \ x \neq y\right\}.$$

If $[f]_{\mathrm{Lip}(\Omega)}$ is finite, then we say that $f$ is *Lipschitz (continuous)* in $\Omega$ and the quantity $[f]_{\mathrm{Lip}(\Omega)}$ is called the Lipschitz constant. If $f$ is Lipschitz continuous in some neighborhood of each point of $\Omega$, then we say that $f$ is *locally Lipschitz* in $\Omega$.

Here and henceforth, we use standard notations for function spaces. Let $C(A, B)$ denote the space of all *continuous* functions (mappings) from one *metric space* $A$ to another metric space $B$. The space $\mathbf{R}^N$ is regarded as a metric space with the metric $d(z, w) = |z - w|$. If $B$ is $\mathbf{R}$, then we simply write $C(A)$ instead of $C(A, \mathbf{R})$. For an *open* set $A$ in $\mathbf{R}^N$ and for $k = 1, 2, \dots$, by $C^k(A, \mathbf{R}^m)$ we mean the space of functions in $C(A, \mathbf{R}^m)$ whose *partial derivatives* up to the $k$th order are continuous in $A$. For a general set $A$ in $\mathbf{R}^N$, by $C^k(A, \mathbf{R}^m)$ we mean the space of functions on $A$ that are extendable as a function of $C^k(U, \mathbf{R}^m)$ for some open set $U$ containing $A$. If a function is in $C^k(A, \mathbf{R}^m)$ for all $k \geq 1$, it is said to be *smooth* in $A$. The set of all smooth functions in $A$ with values in $\mathbf{R}^m$ is denoted by $C^\infty(A, \mathbf{R}^m)$. It is easy to see that a locally Lipschitz function is continuous (but may not be differentiable).

By the fundamental theorem of calculus, it is well known that any $C^1$ function is locally Lipschitz.

We begin by stating uniqueness of a solution when an $\mathbf{R}^N$-valued function (a vector field) $b$ is also continuous in time and (globally) Lipschitz in $\mathbf{R}^N$.

---

**Proposition 1.1**

*Let an $\mathbf{R}^N$-valued function $b$ be continuous in $\mathbf{R}^N \times I$ with $I = [0, a]$, which is a given interval with $a > 0$. Assume that $b$ satisfies the Lipschitz condition of the form*

$$[b(\cdot, t)]_{\mathrm{Lip}(\mathbf{R}^N)} \leq M, \tag{1.2}$$

*with some constant independent of $t \in I$. Let $x_i \in C^1(\mathring{I}, \mathbf{R}^N) \cap C(I, \mathbf{R}^N)$ be a solution of (1.1), with initial datum $x_i(0) = X_i \in \mathbf{R}^N$ for $i = 1, 2$. Then*

$$|x_1(t) - x_2(t)| \leq |X_1 - X_2| \exp(Mt), \quad t \in I. \tag{1.3}$$

*In particular, there is at most one solution $x$ for (1.1) with a given initial datum $x(0) = X \in \mathbf{R}^N$.*

---

Here $\mathring{C}$ or int $C$ denotes the *interior* of a set $C$, i.e., the largest open set contained in $C$. For $I$, $\mathring{I} = (0, a)$. We use standard notation of the *intersection* $V \cap W$ of two sets $V$ and $W$ in a set $Z$

$$V \cap W = \{z \in Z \mid z \in V \text{ and } z \in W\}.$$

The *union* $V \cup W$ is defined as

$$V \cup W = \{z \in Z \mid z \in V \text{ or } z \in W\}.$$

In Proposition 1.1, $Z$ is taken as the space of all $\mathbf{R}^N$-valued functions on $I$ and $V = C^1(\mathring{I}, \mathbf{R}^N)$, $W = C(I, R^N)$ to define $V \cap W$. We also define

$$\bigcup_{\lambda \in \Lambda} V_\lambda = \{z \in Z \mid z \in V_\lambda \text{ for some } \lambda \in \Lambda\},$$

$$\bigcap_{\lambda \in \Lambda} V_\lambda = \{z \in Z \mid z \in V_\lambda \text{ for all } \lambda \in \Lambda\}.$$

We shall often write the unique solution $x$ of (1.1) with initial data $x(0) = X$ by $x = x(t, X)$. Proposition 1.1 follows easily from a simple version of Gronwall inequality in the next section. For more general Gronwall inequalities and their applications, see the book by Dragomir [33].

The existence of a solution $x = x(t, X)$ of (1.1) in Proposition 1.1 in a possibly shorter interval $[0, a']$ for some $a' \in (0, a]$ is standard and known as a local existence theorem. Moreover, the continuity of $b$ alone is enough to guarantee such an existence, which is known as Peano's theorem. See, for example, [52, Chapter 2].

### 1.1.2  Gronwall Inequality

**Lemma 1.2**
*Let $\varphi$ be a nonnegative continuous function in $[0, a]$, with $a > 0$. Assume that*

$$\varphi(t) \le \varphi_0 + \int_0^t M\varphi(s)\mathrm{d}s, \quad t \in [0, a], \tag{1.4}$$

*with some nonnegative constant $\varphi_0$ and $M$. Then*

$$\varphi(t) \le \varphi_0 \exp(Mt), \quad t \in [0, a]. \tag{1.5}$$

***Proof of Proposition 1.1*** Integrating (1.1) on $(0, t)$ with $x = x_i$ for $i = 1, 2$, we see that $x_i$ satisfies the integral equation

$$x_i(t) = X_i + \int_0^t b(x_i(s), s)\,\mathrm{d}s, \quad (i = 1, 2).$$

Thus, the difference $x_1 - x_2$ satisfies

$$x_1(t) - x_2(t) = X_1 - X_2 + \int_0^t (b(x_1(s), s) - b(x_2(s), s))\,\mathrm{d}s. \tag{1.6}$$

We estimate (1.6) and invoke (1.2) to get

$$|x_1(t) - x_2(t)| \le |X_1 - X_2| + \int_0^t M|x_1(s) - x_2(s)|\,\mathrm{d}s. \tag{1.7}$$

We now apply the Gronwall inequality with $\varphi = |x_1 - x_2|$, $\varphi_0 = |X_1 - X_2|$. Since (1.7) is simply (1.4), the estimate (1.5) in Lemma 2.1 yields (1.3). $\qquad\square$

There are several ways to prove Lemma 1.2. We here present two proofs.

***Proof of Lemma 1.2 (Proof by iteration)*** We simply plug the estimate (1.4) into the right-hand side of (1.4) to get

$$\varphi(t) \le \varphi_0 + M \int_0^t \left( \varphi_0 + M \int_0^s \varphi(\tau)d\tau \right) ds$$

$$= (1 + Mt)\varphi_0 + M^2 \int_0^t (t - s)\varphi(s)ds.$$

We repeat this procedure (a total of $m$ times) to get

$$\varphi(t) \le \sum_{j=0}^m \frac{(Mt)^j}{j!}\varphi_0 + \frac{M^{m+1}}{m!} \int_0^t (t - s)^m \varphi(s)ds.$$

We now obtain (1.5) since

$$\left| \frac{M^{m+1}}{m!} \int_0^t (t - s)^m \varphi(s)ds \right| \le \frac{M^{m+1}t^{m+1}}{(m + 1)!} \sup_{s\in(0,t)} \varphi \to 0$$

as $m \to \infty$ and

$$\exp Mt = \sum_{j=0}^\infty (Mt)^j / j!$$

understanding that $0! = 1$. $\qquad\square$

***Proof of Lemma 1.2 (Proof by a differential inequality)*** We set the right-hand side of (1.4) by $y(t)$. Then (1.4) implies that $\dot{y} \le My$. If $\varphi_0 > 0$, then $y > 0$ in $[0, a)$, which yields $\dot{y}/y \le M$. Integrate both sides on $[0, t)$ to get

$$\log y(t) - \log \varphi_0 \le Mt.$$

This implies $y(t) \le \varphi_0 \exp(Mt)$, which yields (1.5) for $\varphi_0 > 0$. In the case $\varphi_0 = 0$, we just let $\varphi_0 \downarrow 0$ (i.e., $\varphi_0 \to 0$, with $\varphi_0 > 0$) in (1.5) to conclude that $\varphi \equiv 0$, which clearly satisfies (1.5) with $\varphi_0 = 0$. $\qquad\square$

▶ **Remark 1.3** Of course, Lemma 1.2 can be generalized in several ways.

**(1)** We may allow $\varphi$ just a nonnegative *integrable* function in $[0, a)$, not necessarily continuous in (1.4).

**(2)** The constant $M$ in (1.4) is allowed to be merely an integrable (and nonnegative) function of $s$ in $(0, a)$, retaining the continuity of $\varphi$. The estimate (1.4) should be replaced by

$$\varphi(t) \le \varphi_0 \exp\left(\int_0^t M(s)ds\right).$$

▶ **Remark 1.4** The assumption of the uniqueness statement of Proposition 1.1 can be weakened in several ways. For example, the uniqueness still holds if (1.2) is replaced by

$$[b(\cdot, t)]_{\mathrm{Lip}(B_R)} \le M(t)$$

for all $B_R$, the closed ball of radius $R < \infty$ centered at the origin, provided that $M(t)$ is integrable in $(0, a)$. The continuity of $b$ in time is not assumed.

### 1.1.3　Osgood Condition

If the Lipschitz condition is not fulfilled, the solution to the initial value problem may not be unique, as the following simple example shows.

**Example 1.5 (Figure 1.1)**
Consider a scalar differential equation $\dot{x} = \sqrt{|x|}$ with initial data $x(0) = 0$. Other than the solution $x \equiv 0$, we see that $x(t) = t|t|/4$ is a solution. The function $b(x) = \sqrt{|x|}$ does not satisfy the Lipschitz condition. In fact, $[b]_{\mathrm{Lip}(B_R)} = \infty$ no matter how small $R > 0$ is.

**Fig. 1.1** Solutions

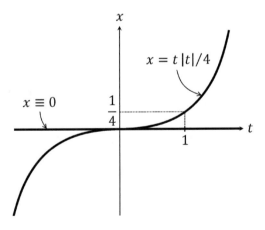

However, the Lipschitz condition is not a necessary condition to guarantee the uniqueness of a solution. We consider a single ordinary equation

$$\dot{x} = f(x), \quad t \in I = [0, a], \tag{1.8}$$

with a continuous function $f$ such that $f(\sigma) > 0$ for $\sigma > 0$, $f(0) = 0$. Assume that the initial data $x(0) = X_0 > 0$. We consider a positive solution in $I$. We divide (1.8) by $f$ to get

$$\dot{x}/f(x) = 1.$$

Integrating over $(0, t)$ we get

$$G(x(t)) - G(X_0) = t$$

if $G$ is a primitive (antiderivative) of $1/f$, for example,

$$G(\sigma) = \int_1^\sigma \frac{d\sigma}{f(\sigma)}.$$

If $\lim_{X_0 \to 0} G(X_0) = -\infty$, in other words,

$$\int_0^\delta \frac{d\sigma}{f(\sigma)} = \infty \tag{1.9}$$

for some $\delta > 0$, there exists no nonnegative solution for (1.8) with $x(0) = 0$ other than the trivial solution $x(t) \equiv 0$ for $t \in I$. Indeed, if there is another solution $\bar{x}$, then there is $t_0 \in [0, a)$ such that $\bar{x}(t) > 0$ for $t \in (t_0, t_0 + \varepsilon)$ and $\bar{x}(t_0) = 0$ for some $\varepsilon > 0$ since $\bar{x}$ is continuous on $I$. Integrating $\dot{x}/f(x) = 1$ over $(t_0 + \varepsilon_1, t)$ for some $\varepsilon_1 \in (0, \varepsilon)$ and $t \in (t_0 + \varepsilon_1, t_0 + \varepsilon)$, we obtain

$$G(\bar{x}(t)) - G(\bar{x}(t_0 + \varepsilon_1)) = t.$$

We fix $t$ and send $\varepsilon_1 \downarrow 0$ in this setting. Since

$$\lim_{X_0 \downarrow 0} G(X_0) = -\infty$$

by (1.9) and $\bar{x}(t_0 + \varepsilon_1) \downarrow 0$ as $\varepsilon_1 \to 0$, we end up with

$$G(\bar{x}(t)) + \infty = t,$$

which is absurd. Thus, $\bar{x}$ cannot be positive in $I$.

If $f(\sigma) = \sigma$, then condition (1.9) is fulfilled. Even if $f(\sigma) = \sigma |\log \sigma|$ so that $f'(0) = \infty$, (1.9) is still fulfilled since

$$\int_0^\delta \frac{d\sigma}{\sigma |\log \sigma|} = \int_{-\infty}^{\log \delta} \frac{d\tau}{|\tau|} = \infty$$

by setting $\tau = \log \sigma$. For later convenience, we say that a continuous nondecreasing function $f$ satisfying $f(\sigma) > 0$ for $\sigma > 0$, $f(0) = 0$ is simply a *modulus*.

**Theorem 1.6**

Let $b$ be continuous in $\mathbf{R}^N \times I$, with $I = [0, a]$, with $a > 0$. Assume that $b$ satisfies the Osgood condition, i.e., there is a modulus $f$ satisfying (1.9) (for some $\delta > 0$) such that

$$|b(x, t) - b(y, t)| \leq f(|x - y|) \tag{1.10}$$

for all $t \in I$, $x, y \in \mathbf{R}^N$. Then

$$G(|x_1(t) - x_2(t)|) \leq t + G(|X_1 - X_2|) \quad for \quad t \in I \tag{1.11}$$

provided that $X_1 \neq X_2$. Here $x_i \in C^1(\mathring{I}, \mathbf{R}^N) \cap C(I, \mathbf{R}^N)$ is a solution of (1.1) with initial data $x_i(0) = X_i$ for $i = 1, 2$. In particular, there is at most one solution $x$ for (1.1) with a given initial datum $x(0) = X \in \mathbf{R}^N$.

Theorem 1.6 follows from another version of the Gronwall inequality given in the next lemma. Condition (1.10) with (1.9) is often called the Osgood condition.

**Lemma 1.7**

Let $\varphi$ be a nonnegative continuous function in $I = [0, a]$ with $a > 0$. Assume that $\varphi_0$ is given nonnegative number and that $\varphi$ satisfies the inequality

$$\varphi \leq \varphi_0 + \int_0^t f(\varphi(s)) \, ds, \quad t \in [0, a], \tag{1.12}$$

with a modulus $f$. Then

$$G(\varphi) \leq G(\varphi_0) + t, \tag{1.13}$$

where $G$ is a primitive of $1/f(\sigma)$ provided that $\varphi > 0$ on $I$. If $\varphi_0 = 0$, then $\varphi \equiv 0$ provided that $f$ satisfies (1.9) for some $\delta > 0$.

***Proof of Lemma 1.7*** The proof is parallel to that using a differential inequality of Lemma 1.2. Assume that $\varepsilon > 0$ on $I$. We set the right-hand side of (1.12) by $y$ and obtain $\dot{y} = f(\varphi) \leq f(y)$, since $0 \leq \varphi \leq y$ and the modulus $f$ is nondecreasing. Since $f(y) > 0$ for $y > 0$, we have $\dot{y}/f(y) \leq 1$. Integrating both sides of $\dot{y}/f(y) \leq 1$ over $(0, t)$, we get (1.13) since $G(\varphi) \leq G(y)$ by the monotonicity of $G$.

We now discuss the case $\varphi_0 = 0$. Suppose that $\varphi$ were identically equal to zero. Then there would exist $t_0 \geq 0$ ($t_0 < a$) such that $\varphi(t) = 0$ for $t \in [0, t_0]$ and $\varphi(t_j) > 0$ and $t_j \downarrow t_0$ (i.e., $t_j \to t_0$ with $t_j > t_0$) as $j \to \infty$. This would imply $y(t) > 0$ for $t > t_0$ and $y(t) = 0$ for $t \leq t_0$. By (1.13) we see that

$$G\left(\varphi(t)\right) \leq G\left(\varphi(t_j)\right) + t$$

for $t > t_j$. If the condition (1.9) is fulfilled, then $G\left(\varphi(t_j)\right) \to G\left(\varphi(t_0)\right) = -\infty$ since $\varphi(t_j) \to 0$ as $j \to \infty$. This yields a contradiction since $G\left(\varphi(t)\right)$ is finite and independent of $j$. We thus conclude that $\varphi \equiv 0$ if $\varphi_0 = 0$. □

***Proof of Theorem 1.6*** We argue in the same way as in the proof of Proposition 1.1 to get (1.6), i.e.,

$$x_1(t) - x_2(t) = X_1 - X_2 + \int_0^t \left(b\left(x_1(s), s\right) - b\left(x_2(s), s\right)\right) ds.$$

We estimate (1.6) and invoke (1.10) to get

$$|x_1(t) - x_2(t)| \leq |X_1 - X_2| + \int_0^t f\left(|x_1(s) - x_2(s)|\right) ds.$$

We now apply Lemma 1.7 with $\varphi = |x_1 - x_2|$ to get

$$G\left(|x_1(t) - x_2(t)|\right) \leq G\left(|X_1 - X_2|\right) + t.$$

The uniqueness under (1.9) follows from the last statement of Lemma 1.7. □

▶ **Remark 1.8** Theorem 1.6 and Proposition 1.1 are still valid when $\mathbf{R}^N$ is replaced by an infinite-dimensional *Banach space*. However, applications to partial differential equations are rather limited because a differential operator is often unbounded in a fixed Banach space; however, see Sect. 1.3.1 for Lipschitz semigroups.

We also note that we can assert the same uniqueness for a negative interval $[-a, 0]$ with $a > 0$, which we call backward uniqueness under the same assumption. Our uniqueness results presented so far are for the forward uniqueness.

### 1.1.4  A Solenoidal Vector Field Having Bounded Vorticity

We shall give a class of examples of vector fields that satisfy the Osgood condition but may not satisfy the Lipschitz condition. This class of examples is important in hydrodynamics.

We consider a *solenoidal* vector field $u = (u^1, \ldots, u^N)$ in $\mathbf{R}^N$. In other words, we assume $\operatorname{div} u = 0$ in $\mathbf{R}^N$, where $\operatorname{div} u = \sum_{i=1}^{N} \partial_i u^i$, $\partial_i = \partial/\partial x_i$ for $u^i = u^i(x)$, $x = (x_1, \ldots, x_N)$. For a function $f$ of $x$, we define its *gradient* $\nabla f$ by $\nabla f = (\partial_1 f, \ldots, \partial_N f)$. We define the *Laplace operator* $\Delta$ by $\Delta f = \operatorname{div} u$, $u = \nabla f$. To derive the necessary formula, for a moment we assume that $u \in C^1(\mathbf{R}^N)$, i.e., $u$ is continuously differentiable, and that the *support* of $u$ is *compact*, i.e., $\operatorname{supp} u$ is included in $B_R$ for some $R > 0$; here, $\operatorname{supp} u$ is the *closure* of the set of $x$ where $u(x) \neq 0$. The space $C_c^1(\mathbf{R}^N)$ denotes the space of all $C^1$ functions such that its support is compact, in other words, the space of all compactly supported $C^1$ functions.

We consider the case $N = 2$ or $3$ to simplify the notation. In hydrodynamics, we call $\omega = \operatorname{curl} u$ the *vorticity* (field) of the velocity vector field $u$, where

$$\operatorname{curl} u = \left( \partial_2 u^3 - \partial_3 u^2, \partial_3 u^1 - \partial_1 u^3, \partial_1 u^2 - \partial_2 u^1 \right), \qquad N = 3,$$

$$\operatorname{curl} u = \partial_1 u^2 - \partial_2 u^1, \qquad\qquad\qquad\qquad\qquad N = 2.$$

We are interested in whether or not $u$ is Lipschitz continuous when $\omega$ is bounded, assuming, for example, $u$ is continuous or, more generally, a Schwartz distribution. It turns out that $b(x, t) := u(x)$ satisfies a kind of Osgood condition (1.10) with (1.9) but not Lipschitz continuous. In other words, a solenoidal velocity field with bounded vorticity is continuous but may not be Lipschitz continuous.

To see this phenomenon, we recall the Biot-Savart law. For $N = 3$, we notice that

$$-\Delta = \operatorname{curl} \operatorname{curl} - \nabla \operatorname{div}, \quad \nabla = (\partial_1, \ldots, \partial_N),$$

so that $-\Delta u = \operatorname{curl} \omega$ if $\operatorname{div} u = 0$. (The Laplace operator here acts on a vector field. More precisely, the $i$th component of $\Delta u$ equals $\Delta u^i$.) Thus, formally,

$$u = (-\Delta)^{-1} \operatorname{curl} \omega = \operatorname{curl}(-\Delta)^{-1} \omega. \tag{1.14}$$

Here $(-\Delta)^{-1} g$ denotes the Newton potential of $g$, i.e., $(-\Delta)^{-1} g = E * g$, where $E$ is the fundamental solution of $-\Delta$ and $*$ denotes the *convolution*, i.e.,

$$(E * g)(x) = \int_{\mathbf{R}^N} E(x - y) g(y) \, dy$$

$$= \int_{-\infty}^{\infty} \cdots \int_{-\infty}^{\infty} E(x - y) g(y) \, dy_1 \cdots dy_N.$$

In the case $N = 3$, $E(x) = 1/(4\pi |x|)$. Thus, relation (1.14) can be written

$$u = \text{curl}(E * \omega) = \mathbf{K} * \omega; \tag{1.15}$$

this relation is often called the *Biot–Savart law*. Here $\mathbf{K} = (K_{ij})_{1 \leq i, j \leq 3}$ is a $3 \times 3$ matrix field when $N = 3$. Its explicit form is

$$\mathbf{K} = \begin{pmatrix} 0 & -\partial_3 E & \partial_2 E \\ \partial_3 E & 0 & -\partial_1 E \\ -\partial_2 E & \partial_1 E & 0 \end{pmatrix}.$$

Each component of $\mathbf{K}$ (denoted by $K$) satisfies

$$|K(x)| \leq C_1/|x|^{N-1}, \quad |\nabla K(x)| \leq C_2/|x|^N \tag{1.16}$$

for all $x \in \mathbf{R}^N \setminus \{0\}$ with $C_1$ and $C_2$ independent of $x$.

In the case $N = 2$, formula (1.14) is still valid where the vorticity $\omega$ is a scalar and the curl $\varphi$ in (1.14) should be interpreted as $\nabla^\perp \varphi (:= (\partial_2 \varphi, -\partial_1 \varphi))$ for a scalar field $\varphi$. This (1.14) yields (1.15) with $\mathbf{K} = \text{curl } E$, $E(x) = -(\log |x|)/2\pi$; here $\mathbf{K}$ is a 2-vector field. In this case, each component of $\mathbf{K}$ also satisfies (1.16). The derivation of the Biot–Savart law presented here is formal but can be justified; see, for example, [45, Chapter 2] where $N = 2$.

The next general lemma in particular shows that a solenoidal velocity field $u$ (with compact support) satisfies the Osgood condition (1.10) with (1.9) if its vorticity $\omega$ is bounded, i.e., $\|\omega\|_\infty < \infty$, where

$$\|\omega\|_\infty = \sup \left\{ |\omega(x)|, x \in \mathbf{R}^N \right\}$$

for a continuous function $\omega$. The norm $\| \cdot \|_\infty$ should be interpreted as an *essential supremum* norm for just a Lebesgue *measurable* function; see Sect. 1.2.5 for a rigorous definition. The notion of support should be generalized for such $\omega$. For a measurable function $g$, its *support* supp $g$ is the smallest closed set $Z$ such that $g = 0$ *almost everywhere* in $\mathbf{R}^N \setminus Z$.

---

**Lemma 1.9**

*Assume that* supp $g$ *is included in* $B_R$ ($R > 1$) *and* $g$ *is bounded and measurable in* $\mathbf{R}^N$. *Let* $K \in C^1 (\mathbf{R}^N \setminus \{0\})$ *satisfy* (1.16) *with some positive constants* $C_1$ *and* $C_2$. *Set* $v = K * g$. *Then* $v$ *satisfies*

$$|v(x) - v(y)| \leq c_N \|g\|_\infty |x - y| (C_2 |\log |x - y|| + L), \quad x, y \in \mathbf{R}^N, \tag{1.17}$$

*with* $L = C_2 \log R + 2C_1$ *and a constant* $c_N$ *depending only on* $N$.

Since $f(\sigma) = \sigma (a_1 |\log \sigma| + a_2)$ satisfies (1.9) for $a_1, a_2 \geq 0$, (1.17) yields the Osgood condition for a vector field $u$. We shall give a proof of Lemma 1.9.

*Proof.* Since

$$|v(x) - v(y)| \leq \|g\|_\infty \int_{B_R} |K(x - z) - K(y - z)| \, dz,$$

it suffices to prove that

$$\int_{B_R} |K(x - z) - K(y - z)| \, dz \leq c_N |x - y| \left(C_2 |\log |x - y|| + L\right), \qquad (1.18)$$

with $c_N$ depending only on $N$. By translation, we may assume that $x + y = 0$, i.e., the middle point between $x$ and $y$ is the origin, so that $x = -y$ and $|x| = \eta/2$.

We set $|x - y| = \eta$ and observe that

$$\int_{B_R} |K(x - z) - K(y - z)| \, dz \leq \int_{B_R \cap \{|z| \geq \eta\}} |K(x - z) - K(y - z)| \, dz$$

$$+ \int_{B_R \cap \{|z| \leq \eta\}} |K(x - z)| \, dz + \int_{B_R \cap \{|z| \leq \eta\}} |K(y - z)| \, dz$$

$$= I + II + III.$$

To estimate $I$, we notice that $|z| \geq \eta$ implies

$$|z - y - t(x - y)| \geq |z| - |tx + (1 - t)y| \geq |z| - \eta/2 \geq |z|/2 \qquad (1.19)$$

for $t \in [0, 1]$ since $|x| = |y| = \eta/2$ (Fig. 1.2). Since

$$|K(x - z) - K(y - z)| = \left| \int_0^1 \langle \nabla K ((x - y)t + y - z) , x - y \rangle dt \right|$$

$$\leq C_2 \int_0^1 \frac{dt}{|(x - y)t + y - z|^N} |x - y|,$$

estimate (1.19) implies

$$I \leq C_2 |x - y| \int_{B_R \cap \{|z| \geq \eta\}} \frac{dz}{(|z|/2)^N} = C_2 2^N \eta \int_{\eta < r < R} \left( \int_{|\sigma| = 1} \frac{1}{r^N} r^{N-1} d\sigma \right) dr$$

$$= C_2 2^N \eta \int_\eta^R \frac{1}{r} dr \left| S^{N-1} \right| = C_2 2^N \left| S^{N-1} \right| \eta \left(\log(R/\eta)\right),$$

**Fig. 1.2** Location of $z$

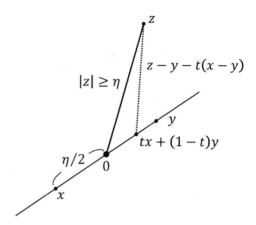

where $\left|S^{N-1}\right|$ denotes the surface area of the unit sphere $S^{N-1}$ in $\mathbf{R}^N$ and $d\sigma$ denotes the surface element. (Here we invoke the *Schwarz inequality* $|\langle x, y\rangle| \leq |x||y|$ at the beginning.) The term $II$ is estimated as

$$II = \int_{B_R \cap \{|z| \leq \eta\}} |K(x-z)| \, dz \leq \int_{\{|z| \leq 2\eta\}} |K(z)| \, dz$$

$$\leq C_1 \int_0^{2\eta} \left( \int_{|\sigma|=1} \frac{1}{r^{N-1}} r^{N-1} d\sigma \right) dr = C_1 \left|S^{N-1}\right| 2\eta.$$

The same estimate holds for $III$. We have thus proved that

$$I + II + III \leq c_N \, \eta \left\{(C_2 \log R + 2C_1) + C_2 |\log \eta|\right\},$$

which yields (1.18), where $c_N = 2^N \left|S^{N-1}\right|$. $\qquad\square$

The function $v$ may not be Lipschitz continuous in Lemma 1.9 even if $g$ is continuous. Moreover, there is an explicit counterexample of a solenoidal vector field $u$ in $\mathbf{R}^2$ whose vorticity $\omega$ is bounded, but $u$ itself is not Lipschitz continuous, and its derivative $\partial_j u^i$ $(1 \leq i, j \leq 2)$ has a logarithmic singularity. In fact, if $\varphi \in C_c^1(\mathbf{R}^2) \cap C^\infty (\mathbf{R}^2 \backslash \{0\})$ satisfies $\varphi(x_1, x_2) = x_1 x_2 \, (-\log |x|)^\alpha$ with $\alpha \in (0, 1)$ near $x = 0$, then $u = \nabla^\perp \varphi$ satisfies $\operatorname{div} u = 0$ and $\operatorname{curl} u = -\Delta \varphi = \omega$ is a continuous function with compact support, so that $\omega$ is bounded. It is easy to prove that $u$ is not Lipschitz continuous (Exercise 1.5). However, by Lemma 1.9, it satisfies the Osgood condition. Thus, there is at most one solution for the initial value problem for $\dot{x} = u(x)$.

### 1.1.5  Equation with Fractional Time Derivative

We next discuss the uniqueness problem for equations involving fractional time derivatives. A diffusion equation with a fractional time derivative is an interesting topic since it describes various interesting physical phenomena like anomalous diffusion. We here consider the case when it only depends on time. We recall a Caputo derivative with $\alpha \in (0, 1)$ starting from $t = 0$, i.e.,

$$\left(\partial_t^\alpha x\right)(t) = \frac{1}{\Gamma(1 - \alpha)} \int_0^t (t - s)^{-\alpha} \partial_t x(s) ds, \tag{1.20}$$

where $\Gamma(\beta)$ denotes the Gamma function, i.e.,

$$\Gamma(\beta) = \int_0^\infty e^{-s} s^{\beta - 1} ds.$$

Instead of (1.1), we consider

$$\partial_t^\alpha x = b(x, t), \tag{1.21}$$

where $b = b(\ ,t)$ is a time-dependent vector field in $\mathbf{R}^N$ and $t > 0$. We give a uniqueness result similar to Proposition 1.1.

> **Proposition 1.10**
> *Assume that $\alpha \in (0, 1)$. Let $b$ be continuous in $\mathbf{R}^N \times I$, with $I = [0, a)$, which is a given interval with $a \in (0, \infty]$. Assume that $b$ fulfills (1.2). Then the solution $x \in C^1(\mathring{I}, \mathbf{R}^N) \cap C(I, \mathbf{R}^N)$ of (1.21) in $\mathring{I}$ with a given initial datum $x(0) = X$ is unique provided that $\int_I |\dot{x}| dt < \infty$.*

We do not state the estimate of $x(t, X_1) - x(t, X_2)$ here. To prove this proposition, we need a slightly different type of Gronwall inequality. We just give the simplest version.

> **Lemma 1.11**
> *Let $\varphi$ be a nonnegative continuous function in $[0, a]$, with $a > 0$. Assume that*
>
> $$\varphi(t) \le M_0 \int_0^t (t - s)^{-\beta} \varphi(s) ds, \quad t \in [0, a], \tag{1.22}$$
>
> *with some constant $M_0 \ge 0$, $\beta \in (0, 1)$. Then $\varphi \equiv 0$.*

We give here a proof by iteration since another one based on differential inequality is more involved.

***Proof of Lemma 1.11*** Estimate (1.22) yields

$$0 \le \varphi(t) \le M_0 \|\varphi\|_\infty t^{1-\beta}/(1-\beta).$$

Plug this inequality into the right-hand side of (1.22) to get

$$0 \le \varphi(t) \le \frac{M_0^2 \|\varphi\|_\infty}{1-\beta} \frac{\Gamma(1-\beta)\Gamma(2-\beta)}{\Gamma(3-2\beta)} t^{2-2\beta}.$$

Here we notice that

$$\int_0^t (t-s)^{p-1} s^{q-1} ds = t^{p+q-1} B(p,q),$$

$$B(p,q) = \int_0^1 (1-\tau)^{p-1} \tau^{q-1} d\tau,$$

and we invoke the elementary formula indicating the relation with the beta and gamma functions

$$B(p,q) = \frac{\Gamma(p)\Gamma(q)}{\Gamma(p+q)}.$$

Repeat the argument to get

$$0 \le \varphi(t) \le \frac{M_0^k C}{1-\beta} \frac{\Gamma(1-\beta)^{k-1}\Gamma(2-\beta)}{\Gamma(k+1-k\beta)} t^{k(1-\beta)}, \quad k = 1, 2, \ldots.$$

The right-hand side is dominated by

$$M_1^k M_2 / \Gamma(k+1-k\beta)$$

for $t \in [0, a]$ by taking constants $M_1, M_2$ in a suitable way (which is independent of $k$). Since $\beta \in (0, 1)$, the asymptotic behavior of $\Gamma(k+1-k\beta)$ as $k \to \infty$ (Stirling's formula) yields that $M_1^k M_2 / \Gamma(k+1-k\beta) \to 0$ as $k \to \infty$. Thus, we conclude that $\varphi \equiv 0$. $\qquad\square$

**Proof of Proposition 1.10** For $\gamma > 0$, let $\mathbf{I}^\gamma$ denote the Riemann–Liouville operator, i.e.,

$$(\mathbf{I}^\gamma g)(t) = \frac{1}{\Gamma(\gamma)} \int_0^t (t - s)^{\gamma - 1} g(s) \, ds$$

$$= \frac{1}{\Gamma(\gamma)} t_+^{\gamma - 1} * g, \quad t_+ = \max(t, 0).$$

By direct computation, we easily see that $\mathbf{I}^{\gamma_1 + \gamma_2} = \mathbf{I}^{\gamma_1} \mathbf{I}^{\gamma_2}$ for $\gamma_1, \gamma_2 > 0$. Since

$$\partial_t^\alpha x = \mathbf{I}^{1-\alpha} \dot{x},$$

we observe that $\mathbf{I}^\alpha \partial_t^\alpha x = \mathbf{I}^1 \dot{x} = x(t) - x(0)$. Here, we invoked the assumption $\int_I |\dot{x}| \, dt < \infty$ to guarantee the integrability of the integrand in $\mathbf{I}^{1-\alpha} \dot{x}$ near $s = 0$, so that $(\mathbf{I}^{1-\alpha} \dot{x})(t)$ is finite for $t \in I$. Applying $\mathbf{I}^\alpha$ to both sides of (1.21) we get

$$x(t) = X + \mathbf{I}^\alpha b. \tag{1.23}$$

Let $x_1$ and $x_2$ be two solutions of (1.23). Then the difference satisfies

$$x_1(t) - x_2(t) = \mathbf{I}^\alpha \left( b(x_1, t) - b(x_2, t) \right).$$

We set $\varphi = |x_1(t) - x_2(t)|$ and estimate the preceding inequality using (1.2) to get

$$\varphi \leq \mathbf{I}^\alpha M \varphi \quad \text{in} \quad [0, a].$$

We are now in a position to apply Lemma 1.11 with $\beta = 1 - \alpha$, $M_0 = M / \Gamma(\alpha)$ to conclude that $\varphi \equiv 0$. □

▶ **Remark 1.12** Unlike (1.1), the solution of (1.21) may be nondifferentiable at $t = 0$, even if $b$ is smooth. Indeed, if we consider $x = t^\alpha$, then $x$ is not differentiable at $t = 0$ since $\alpha \in (0, 1)$. A direct calculation shows that $\partial_t^\alpha x = \Gamma(\alpha + 1)$, $x(0) = 0$.

## 1.2   Gradient Flow of a Convex Function

As we observed in Example 1.5, the continuity of $b$ in (1.1) is not enough to guarantee the uniqueness of the solution for its initial value problem. On the other hand, the existence of a local-in-time solution is known as Peano's existence theorem. If $b$ loses the continuity in $x$, we expect to have neither existence nor uniqueness in general. However, if $b(x)$ is a "monotone" decreasing function not necessarily continuous, we expect to have uniqueness for positive time. Such a situation typically arises when $b$ is the negative of the gradient of a convex function. Since the argument still works in a Hilbert space, not just in $\mathbf{R}^N$, we shall discuss

(1.2) where $x$ has values in a Hilbert space. This has wide application to partial differential equations, which are regarded as the gradient flow of a convex function.

### 1.2.1  Maximal Monotone Operator and Unique Existence of Solutions

Let $H$ be a real *Hilbert space* equipped with an inner product $\langle \, , \, \rangle$. The norm $\|x\|$ of $x$ in $H$ is defined by $\|x\| = \langle x, x \rangle^{1/2}$. Let $A$ be a mapping from a given subset $D(A)$ of $H$ to $2^H \backslash \{\emptyset\}$, the set of all nonempty subsets of $H$. The mapping $A$ is often called a set-valued operator or a multivalued function. We say that $A$ is *monotone* if

$$\langle x' - y', x - y \rangle \geq 0$$

for all $x, y \in D(A)$ and $x' \in Ax$, $y' \in Ay$, where $Ax$ denotes the set $A(x)$ to simplify the notation (Fig. 1.3). The set $D(A)$ is often called the *domain* of $A$. Here is a trivial example. The space $\mathbf{R}^N$ is, of course, a real (finite-dimensional) Hilbert space equipped with a standard inner product

$$\langle x, y \rangle = \sum_{i=1}^{N} x_i y_i$$

for $x = (x_1, \ldots, x_N)$, $y = (y_1, \ldots, y_N)$. In the case $H = \mathbf{R}$, $Ax = \{f(x)\}$ is monotone if and only if $f$ is monotone nondecreasing.

**Example 1.13 (Figure 1.4)**
Let $H = \mathbf{R}$, and set $D(A) = H$ and $Ax = \{+1\}$ for $x > 0$, $Ax = \{-1\}$ for $x < 0$. $A0$ is a given subset in $[-1, 1]$. This operator is a monotone operator in $\mathbf{R}$.

The reason we allow "multivalued" is that the notion of a maximal monotone operator is important. We say that $A \subset B$ for two set-valued operators $A$ and $B$ in $H$ if $D(A) \subset D(B)$ and $Ax \subset Bx$ for all $x \in D(A)$. By $U \subset W$ we mean that $U$ is *included* in $W$, i.e., an element of $U$ is always an element of $W$. We say that a monotone operator $A$ is *maximal* if $A$ is maximal for the foregoing inclusion. In other words, if there is a monotone operator $B \supset A$, then operator $A$ agrees with $B$. One observes that in Example 1.13, operator $A$ is a maximal monotone operator if and only if $A0 = [-1, 1]$.

**Fig. 1.3** Monotonicity

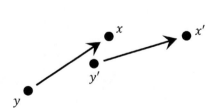

**Fig. 1.4** The graph of
operator $A$

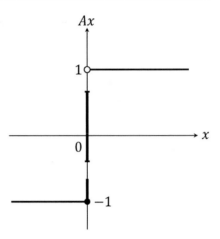

We shall consider an ordinary differential equation (inclusion) for $x = x(t)$ in $H$
of the form

$$\dot{x} \in -Ax, \tag{1.24}$$

where $-Ax = \{-y \mid y \in Ax\}$. Here is a fundamental unique existence theorem due
to Kōmura [65].

**Theorem 1.14**
*Let $A$ be a maximal monotone operator in a real Hilbert space $H$. Assume
that $X \in D(A)$. Then there exists a unique $x \in C([0, \infty), H)$ that is Lipschitz
continuous in $[0, \infty)$ such that $x$ solves (1.24) with $x(0) = X$ for almost every
$t > 0$.*

This is rather surprising because it asserts not only the existence of a global solution but also its (forward) uniqueness, i.e., uniqueness for $t \geq 0$ (cf. Remark 1.8).
The Eq. (1.24) looks ambiguous because $Ax$ is a set but the solutions "know" how
to evolve.

We do not intend to give a proof of its existence. Rather, in what follows, we give
a proof for its uniqueness as a result of the contraction property.

**Theorem 1.15 (Stability and uniqueness)**
*Let $A$ be a monotone operator in $H$. Let $x(t, X_i)$ be a solution of (1.24) with
initial datum $X_i \in D(A)$, $i = 1, 2$, in the sense of Theorem 1.14. Then*

$$\|x(t, X_1) - x(t, X_2)\| \leq \|X_1 - X_2\| \quad \text{for all} \quad t \geq 0.$$

(continued)

**Theorem 1.15** (continued)
*In other words, the mapping $X \mapsto x(t, X)$ for $t \geq 0$ (often called a flow map) is a contraction, and in particular, $x(t, X)$ is uniquely determined by $X$.*

*Proof.* We set $x_i = x(t, X_i)$. The difference solves

$$\frac{d}{dt}(x_1 - x_2) = -(y_1 - y_2) \quad \text{for a.e.} \quad t > 0,$$

with $y_i \in Ax_i$. Take the inner product of both sides with $x_1 - x_2$ to get

$$\left\langle x_1 - x_2, \frac{d}{dt}(x_1 - x_2) \right\rangle = -\langle x_1 - x_2, y_1 - y_2 \rangle.$$

Since $A$ is monotone, the right-hand side is always nonpositive, so that

$$\frac{1}{2}\frac{d}{dt}\|x_1 - x_2\|^2 = \left\langle x_1 - x_2, \frac{d}{dt}(x_1 - x_2) \right\rangle \leq 0.$$

This yields the desired contraction property. In particular, $x(t, X)$ is uniquely determined by taking $X_1 = X_2$. (In this proof we do not invoke the maximality of $A$. The maximality is invoked to prove the existence of a solution.)    $\square$

By uniqueness the flow map $x(t, X)$ possesses the semigroup property $x(t + s, X) = x(t, x(s, X))$, $t, s \geq 0$, so this flow map is often called a contraction semigroup.

### 1.2.2   Canonical Restriction

We are interested in the velocity $\dot{x}$ because it looks ambiguous from (1.24). To characterize this, we recall the notion of a minimal section or canonical restriction.

For a given $x \in D(A)$, we take $y_x \in Ax$ satisfying

$$\|y_x\| = \inf\{\|y\| \mid y \in Ax\}.$$

This is an element of $Ax$ closest to the origin. Here, $\inf S$ for a set $S$ in $\mathbf{R}$ denotes the *infimum* of a set $S$. If $A$ is maximal monotone, it is easy to see that $Ax$ is a nontrivial closed convex set,[1] so there exists $y_x$ uniquely. The mapping $x \mapsto y_x$ is called the

---

[1] A set $C$ in a (real) vector space is said to be *convex* if the segment between arbitrary two points $x, y \in C$ is always cotained in $C$.

*canonical restriction* (or the minimal section) of $A$. This operator is often denoted by $A^0$, i.e., $y_x = A^0 x$.

---

**Theorem 1.16**
*Let $x = x(t, X)$ be the solution of (1.24), with $x(0, X) = X$ given in Theorem 1.14. Then it is right differentiable in time for all $t \geq 0$. Moreover,*

$$\frac{d^+}{dt} x = -A^0 x$$

*for all $t \geq 0$, where $d^+ x / dt$ denotes the right derivative.*

---

We just provide a formal proof assuming that $d^+ x / dt$ exists at $t_0$ and $dx/dt$ is right continuous at $t_0$.

Since $dx/dt \in -Ax$, we see, by the monotonicity of $A$, that

$$\left\langle \frac{dx}{dt}(t_0 + s) - (-\zeta), \ x(t_0 + s) - x(t_0) \right\rangle \leq 0 \quad (s > 0)$$

for any $\zeta \in Ax(t_0)$. Dividing by $s > 0$ and letting $s \downarrow 0$, we see that

$$\left\langle \frac{dx}{dt}(t_0), \ \frac{dx}{dt}(t_0) \right\rangle + \left\langle \zeta, \ \frac{dx}{dt}(t_0) \right\rangle \leq 0.$$

Thus, by the *Schwarz inequality* ($|\langle x, y \rangle| \leq \|x\| \|y\|$ for all $x, y \in H$), we have

$$\left\| \frac{dx}{dt}(t_0) \right\|^2 \leq \|\zeta\| \left\| \frac{dx}{dt}(t_0) \right\|.$$

This implies

$$\left\| \frac{dx}{dt}(t_0) \right\| \leq \|\zeta\|.$$

We thus observe that $-\dfrac{dx}{dt}(t_0) = A^0 x(t_0)$.

We conclude this subsection by studying an implicit Euler scheme for (1.24) since it provides a method to deduce the canonical restriction. The scheme is of the form

$$\begin{cases} \dfrac{x^{k+1} - x^k}{\tau} \in -Ax^{k+1} & k = 1, 2, \ldots, \\ x^1 = X, \end{cases}$$

where $\tau > 0$ is a time grid size that is usually small. In each step, we must solve the resolvent equation

$$x - x_0 \in -\tau Ax \quad \text{or} \quad x + \tau Ax \ni x_0$$

for a given $x_0 \in H$. Here, as previously, we write

$$-\tau Ax = \{-\tau y \mid y \in Ax\} \quad \text{and} \quad x + \tau Ax = \{x + \tau y \mid y \in Ax\}.$$

The last operator $x \mapsto x + \tau Ax$ is often written $I + \tau A$, where $I$ denotes the identity operator.

The *range (image)* $R(I + \tau A)$ of the operator $I + \tau A$ is defined by

$$R(I + \tau A) = \{x + \tau y \mid y \in Ax, \ x \in D(A)\}.$$

**Lemma 1.17**
*Let $A$ be a maximal monotone operator. Let $x_\tau$ be the solution of the resolvent equation $x + \tau Ax \ni x_0$ for $x_0 \in D(A)$ and $\tau > 0$. Then $(x_\tau - x_0)/\tau$ converges to $-A^0 x_0$ strongly in $H$ as $\tau \downarrow 0$.*

▶ **Remark 1.18**

(i) Minty's theorem [73] states that a monotone operator is maximal if and only any of $x + \tau Ax \ni x_0$ has a solution $x$ for any $x_0 \in H$ and any $\tau > 0$. One may relax the condition for $\tau$ just for some $\tau > 0$. The proof of the existence of a solution from maximality is nontrivial, and Hilbert space structure plays a key role. Indeed, there is a counterexample for a Banach space like $\ell^p$ ($1 < p < \infty$, $p \neq 2$) [27]; see also [74, Example 2.6].
It is easy to see that solution $x_\tau$ is unique. The proof of the uniqueness is parallel to that of Theorem 1.15.

(ii) In the literature, the solvability of

$$x + \tau Ax = x_0 \quad \text{for arbitrary} \quad x_0 \in H$$

is often written $R(I + \tau A) = H$. The solution operator from $x_0 \mapsto x_\tau$ is called the resolvent operator (often denoted by $J_\tau$). The operator $J_\tau$ is called the Yosida operator.

(iii) Since $(x_\tau - x_0)/\tau = (J_\tau x_0 - x_0)/\tau = -A J_\tau x_0$, the convergence result says that $A J_\tau x_0 \to A^0 x_0$ if $x_0 \in D(A)$. The operator $A_\tau = A J_\tau$ is called the Yosida approximation of $A$.

**(iv)** Although $x_\tau = J_\tau x_0$ is defined for any $x_0 \in H$, the convergence $x_\tau \to x_0$ in $H$ as $\tau \to 0$ does not hold for general $x_0$. For example, consider $Ax = 0$ for $x \in (-1, 1)$ $A(-1) = (-\infty, 0]$, $A(+1) = [0, \infty)$ for $x \in H = \mathbf{R}$. Then $x_\tau$ does not converge to $x_0$ if $x_0 \notin [-1, 1]$.

*Proof of Lemma 1.17 (based on properties of the weak convergence)* We consider the equation

$$x_\tau + \tau A x_\tau \ni x_0, \quad \tau > 0.$$

We first establish a bound for a difference quotient $\zeta_\tau = (x_\tau - x_0)/\tau$, $\tau > 0$. We estimate the norm of $\zeta_\tau$ to get

$$\|\zeta_\tau\|^2 = \left\langle -Ax_\tau, \frac{x_\tau - x_0}{\tau} \right\rangle$$

$$= -\left\langle Ax_\tau - \eta_0, \frac{x_\tau - x_0}{\tau} \right\rangle - \left\langle \eta_0, \frac{x_\tau - x_0}{\tau} \right\rangle, \quad \eta_0 \in Ax_0$$

$$= -\frac{1}{\tau}\langle Ax_\tau - \eta_0, x_\tau - x_0 \rangle - \langle \eta_0, \zeta_\tau \rangle$$

$$\leq -\langle \eta_0, \zeta_\tau \rangle$$

by the monotonicity of $A$. Thus,

$$\|\zeta_\tau\|^2 \leq \|A^0 x_0\| \|\zeta_\tau\|$$

by the Schwarz inequality. This yields

$$\|\zeta_\tau\| \leq \|A^0 x_0\| \quad \text{or} \quad \|A_\tau x_0\| \leq \|A^0 x_0\| \tag{1.25}$$

using the notation in Remark 1.18 (iii). In particular, we have $x_\tau \to x_0$ in $H$ as $\tau \downarrow 0$.

Because of this bound, $\zeta_{\tau_k}$ weakly converges to some $\zeta \in H$ by taking a subsequence as $\tau_k \downarrow 0$ ($k \to \infty$) if necessary. Here, we invoked a weak compactness of a bounded set in a Hilbert space, that is, if $\|y_m\|$ is bounded, then there is a subsequence $y_{m_k}$ and $y$ such that

$$\langle y_{m_k}, z \rangle \to \langle y, z \rangle \quad \text{for all} \quad z \in H$$

as $k \to \infty$, i.e., $y_{m_k}$ converges to $y$ weakly in $H$, which is denoted by $y_{m_k} \rightharpoonup y$. For the proof, see [19, Theorem 3.16] and [90]. Since $A$ is maximal,

$$x_{\tau_k} \to x_0, \quad \zeta_{\tau_k} \rightharpoonup \zeta \text{ (weakly)}, \quad -\zeta_{\tau_k} \in Ax_{\tau_k}$$

(as $k \to \infty$) implies that

$$- \zeta \in A x_0.$$

Indeed, $\langle -\zeta_{\tau_k} - \eta, \; x_{\tau_k} - u \rangle \to \langle -\zeta - \eta, \; x_0 - u \rangle$ for any $u, \eta \in H$ as $\tau_k \downarrow 0$ so that if $\eta \in Au$, then $\langle -\zeta - \eta, \; x_0 - u \rangle \geq 0$ since $\langle -\zeta_\tau - \eta, x_\tau - u \rangle \geq 0$ by maximality of $A$. If $-\zeta \notin A x_0$, this would contradict the maximality of $A$. Thus, we conclude that $-\zeta \in A x_0$.

By (1.25) we see that $-\zeta = A^0 x_0$. Since any subsequence of $\{\zeta_\tau\}$ admits a weak convergent subsequence with the limit $\zeta$ independently of the choice of a subsequence, $\zeta_\tau$ weakly converges to $\zeta$ without taking a subsequence as $\tau \downarrow 0$. Since $\zeta_\tau \rightharpoonup \zeta$, we see that

$$\|\zeta\| = \sup_{\|\varphi\|=1} \langle \zeta, \varphi \rangle = \sup_{\|\varphi\|=1} \lim_{\tau \downarrow 0} \langle \zeta_\tau, \varphi \rangle \leq \liminf_{\tau \downarrow 0} \sup_{\|\varphi\|=1} \langle \zeta_\tau, \varphi \rangle = \liminf_{\tau \downarrow 0} \|\zeta_\tau\|.$$

(This property is known as a lower semicontinuity of norm under weak convergence; see [19, Proposition 3.5] and [90].) Since $\|\zeta_\tau\| \leq \|\zeta\|$, we observe that $\|\zeta_\tau\| \to \|\zeta\|$ as $\tau \downarrow 0$. Since $\zeta_\tau \rightharpoonup \zeta$, this now implies the strong convergence of $\zeta_\tau$ to $\zeta$. Indeed,

$$\|\zeta_\tau - \zeta\|^2 = \|\zeta_\tau\|^2 + \|\zeta\|^2 - 2\langle \zeta_\tau, \zeta \rangle \to \|\zeta\|^2 + \|\zeta\|^2 - 2\|\zeta\|^2 = 0$$

as $\tau \downarrow 0$.                                                                                      □

### 1.2.3  Subdifferentials of Convex Functions

A typical example of a maximal monotone operator is the subdifferential of a lower semicontinuous convex function. We first recall the definition of a subdifferential. Let $E$ be a function in $H$ with values in $\mathbf{R} \cup \{+\infty\}$. (If $H$ is of infinite dimension, we often call it a functional instead of a function.) We say that $\zeta \in H$ is a *subgradient* of $E$ at $x \in H$ if $E(x) < \infty$ and

$$E(x + h) - E(x) \geq \langle h, \zeta \rangle$$

holds for all $h \in H$. The set of all subgradients of $E$ at $x$ is denoted by $\partial E(x)$. The operator $\partial E$, which corresponds $x$ to the set of all subgradients of $E$ at $x$, is called a *subdifferential* operator but is often called a subdifferential of $E$. The domain $D(\partial E)$ of the subdifferential is defined as

$$D(\partial E) = \{ x \in H \mid \partial E(x) \text{ is not empty} \}.$$

We are interested in the case where $E$ is *convex*, i.e.,

$$E\left(\lambda x + (1 - \lambda)y\right) \leq \lambda E(x) + (1 - \lambda)E(y), \text{ for all } \lambda \in [0, 1], \quad x, y \in H$$

with the interpretation that $\infty + a = \infty$, $a \leq \infty$ for any $a \in \mathbf{R} \cup \{\infty\}$. By definition, the domain

$$D(E) = \{x \in H \mid E(x) < \infty\}$$

must be convex. By definition, $D(\partial E) \subset D(E)$.

A simple example with a multivalued subdifferential is the subdifferential of a convex function $E(x) = |x|$ when $H = \mathbf{R}$. In this case, $\partial E$ is simply $A$ of Example 1.13, with $A0 = [-1, 1]$.

We say that $E$ is *lower semicontinuous* if

$$E(x) \leq \liminf_{y \to x} E(y), \quad x \in H.$$

An *indicator function* $I_K$ of a set $K$ in $H$ defined by

$$I_K(x) = \begin{cases} 0, & x \in K, \\ \infty, & x \in H \setminus K, \end{cases}$$

is lower semicontinuous if and only if $K$ is closed. The function $I_K$ is convex if and only if $K$ is convex. See Figs. 1.5 and 1.6 when $H = \mathbf{R}$.

**Fig. 1.5** Convex and lower semicontinuous

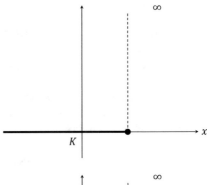

**Fig. 1.6** Convex but not lower semicontinuous

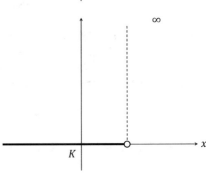

**Lemma 1.19**

*Let $E$ be a convex, lower semicontinuous function in a Hilbert space $H$ with values in $\mathbf{R} \cup \{+\infty\}$. Assume that $E \not\equiv \infty$. Then $A = \partial E$ is a maximal monotone operator.*

***Proof (Monotonicity)*** We take $\zeta_i \in \partial E(x_i)$ for $x_i$ $(i = 1, 2)$. By definition, we have

$$E(x_1 + h) - E(x_1) \geq \langle \zeta_1, h \rangle, \quad E(x_2) - E(x_2 - h) \leq \langle \zeta_2, h \rangle.$$

Take $h = x_2 - x_1$ and subtract the second inequality from the first. This yields

$$\langle \zeta_1 - \zeta_2, x_2 - x_1 \rangle \leq 0,$$

which is the desired monotonicity.

*(Maximality).* By Minty's theorem (Remark 1.18 (i), Exercise 1.7), it suffices to prove that for any $x_0 \in H$ there is a (unique) solution $x = x_*$ of the equation

$$x + \tau \partial E(x) \ni x_0$$

or

$$\frac{1}{\tau}(x - x_0) + \partial E(x) \ni 0$$

for some $\tau > 0$. The left-hand side equals $\partial E_\tau(x)$ if we set

$$E_\tau(x) = \frac{\|x - x_0\|^2}{2\tau} + E(x)$$

(see Exercise 1.10). Thus, $x + \tau \partial E(x) \ni x_0$ is equivalent to $\partial E_\tau(x) \ni 0$. If we are able to find a minimizer $x_*$ of $E_\tau$, then we have $E_\tau(x_* + h) \geq E_\tau(x_*)$ for all $h \in H$. So we have $E_\tau(x_* + h) - E_\tau(x_*) \geq \langle 0, h \rangle$ for all $h \in H$, so that $\partial E_\tau(x_*) \ni 0$. Thus, it suffices to find a minimizer $x_*$ of $E_\tau$.

We shall prove that there exists a minimizer $x_*$ of $E_\tau$ for a fixed $\tau > 0$ by what is called a direct method. Since $E$ is convex and lower semicontinuous, $E(x) + \langle a, x \rangle$ is bounded from below (i.e., $\inf (E(x) + \langle a, x \rangle) > -\infty$) with some $a \in H$ [19, Proposition 1.10]. This looks easy, but it is nontrivial and follows from the Hahn–Banach theorem on extension of linear functionals [19, Chapter 1], [90, Chapter IV]. We admit this fact and prove a kind of coercivity: $\lim_{\|x\| \to \infty} E_\tau(x) = \infty$. In fact, we shall prove a stronger result of the form

$$E_\tau(x) \geq \frac{1}{4\tau}\|x - x_0\|^2 - C \quad \text{for} \quad x \in H$$

with some constant $C > 0$ independent of $x$. By the Schwarz inequality,

$$|\langle a, x - x_0 \rangle| \leq \|a\| \|x - x_0\| = \left( (2\tau)^{1/2} \|a\| \right) \left( (2\tau)^{-1/2} \|x - x_0\| \right).$$

Since $\alpha\beta \leq (\alpha^2 + \beta^2) / 2$ for $\alpha, \beta \geq 0$, we conclude that

$$|\langle a, x - x_0 \rangle| \leq \frac{1}{4\tau} \|x - x_0\|^2 + \tau \|a\|^2.$$

Thus,

$$|\langle a, x \rangle| \leq |\langle a, x - x_0 \rangle| + |\langle a, x_0 \rangle|$$

$$\leq \frac{1}{4\tau} \|x - x_0\|^2 + C_1$$

with $C_1 = |\langle a, x_0 \rangle| + \tau \|a\|^2$. (This says that a linear function is estimated from above by $\delta \|x - x_0\|^2$ for sufficiently large $x$ for any choice of $\delta > 0$.) Thus, we observe that

$$E_\tau(x) = \frac{1}{2\tau} \|x - x_0\|^2 - \langle a, x \rangle + E(x) + \langle a, x \rangle$$

$$\geq \frac{1}{4\tau} \|x - x_0\|^2 - C,$$

with $C = C_1 - \inf (E(x) + \langle a, x \rangle)$. Hence, we obtain $\lim_{\|x\| \to \infty} E_\tau(x) = \infty$.

We now prove the existence of a minimizer of $E_\tau$. Let $\{x_j\}_{j=1}^\infty$ be a minimizing sequence of $E_\tau$, i.e., $\{x_j\}_{j=1}^\infty$ satisfies

$$\lim_{j \to \infty} E_\tau(x_j) = \inf E_\tau := \inf \left\{ E_\tau(x) \mid x \in H \right\}.$$

(Our assumption $E \not\equiv \infty$ evidently implies $\inf E_\tau < \infty$.) By the definition of $\{x_j\}_{j=1}^\infty$, there is a constant $M$ such that $E_\tau(x_j) \leq M$ for all $j \geq 1$. Since $\lim_{\|x\| \to \infty} E_\tau(x) = \infty$, we conclude that $\{x_j\}_{j=1}^\infty$ is a bounded sequence in $H$. Since $\{x_j\}_{j=1}^\infty$ is a bounded sequence, as in the proof of Lemma 1.17, $\{x_j\}$ has a subsequence $\{x_{j_\ell}\}$ converging *weakly* to some $x_* \in H$. In other words, $x_{j_\ell} \rightharpoonup x_*$ as $\ell \to \infty$. We shall prove that $x_*$ is indeed a minimizer of $E_\tau$. We now recall Mazur's theorem [19, Corollary 3.8] (see also [90]), which asserts that if $z_\ell \rightharpoonup z$ (weakly), then some convex combination of $\{z_\ell\}_{\ell=1}^\infty$ converges to $z$ strongly in $H$. We apply Mazur's theorem to conclude that for each $k = 0, 1, 2, \ldots$, there is a convex combination $\left\{ y_m^{(k)} \right\}$ of $\{x_{j_\ell}\}_{\ell=k}^\infty$ such that $y_m^{(k)}$ converges to $x_*$ strongly in $H$

as $m \to \infty$. In other words, there is an integer $n_m^{(k)} \geq k$ and real numbers $\left\{ \lambda_{j_\ell}^{k,m} \right\}_{\ell=k}^{n_m^{(k)}}$ satisfying

$$\sum_{\ell=k}^{n_m^{(k)}} \lambda_{j_\ell}^{k,m} = 1 \quad \text{and} \quad \lambda_{j_\ell}^{k,m} \geq 0 \quad \text{for} \quad k \leq \ell \leq n_m^{(k)},$$

$$y_m^{(k)} = \sum_{\ell=k}^{n_m^{(k)}} \lambda_{j_\ell}^{k,m} x_{j_\ell}$$

such that $y_m^{(k)} \to x_*$ as $m \to \infty$. (Note that Mazur's theorem is another application of the Hahn–Banach theorem.) By a diagonal argument, there is $m(k)$ such that $y_k = y_{m(k)}^k$ converges to $x_*$ as $k \to \infty$. By definition,

$$y_k = \sum_{\ell=k}^{n_k} \lambda_{j_\ell}^k x_{j_\ell} \quad \text{with} \quad \lambda_{j_\ell}^k = \lambda_{j_\ell}^{k,m(k)}, \quad n_k = n_m^{(k)}.$$

In other words, $y_k$ is a convex combination of $\left\{ x_{j_\ell} \right\}_{\ell=k}^{\infty}$. Moreover, $\| y_k - x_* \| \to 0$ as $k \to \infty$.

Since $E_\tau$ is convex, $\{y_k\}$ is still a minimizing sequence. Indeed, by convexity,

$$E_\tau(y_k) = E_\tau \left( \sum_{\ell=k}^{n_k} \lambda_{j_\ell}^k x_{j_\ell} \right) \leq \sum_{\ell=k}^{n_k} \lambda_{j_\ell}^k E(x_{j_\ell}).$$

Since $\{x_{j_\ell}\}$ is a minimizing sequence of $E_\tau$, for any $\delta > 0$ there is $k(\delta)$ such that $E_\tau(x_{j_\ell}) \leq \inf E_\tau + \delta$ for $\ell \geq k(\delta)$. Thus, for $k \geq k(\delta)$,

$$E_\tau(y_k) \leq \sum_{\ell=k}^{n_k} \lambda_{j_\ell}^k (\inf E_\tau + \delta) = \inf E_\tau + \delta$$

since $\sum_{\ell=k}^{n_k} \lambda_{j_\ell}^k = 1$. Thus, $\{y_k\}$ is a minimizing sequence of $E_\tau$. By the lower semicontinuity of $E_\tau$, we see that

$$E_\tau(x_*) \leq \liminf_{k \to \infty} E_\tau(y_k) = \inf E_\tau.$$

Thus, $x_*$ is a minimizer of $E_\tau$. We have thus proved the existence of a minimizer of $E_\tau$, and the proof is now complete.                                                   $\square$

▶ **Remark 1.20**

**(i)** By Lemma 1.19, we are able to apply Theorems 1.14–1.16 when $A$ is the subdifferential $\partial E$ of a convex, lower semicontinuous function $E$ ($E \not\equiv \infty$). Moreover, the initial datum $X$ is allowed to be in $\overline{D(\partial E)}$, the closure of the domain of $\partial E$ in $H$ with necessary modifications. For example, the modified statement in Theorem 1.14 reads as follows. In the case $A = \partial E$, for any $X \in H$ there exists a unique $x \in C([0, \infty), H)$, which is Lipschitz continuous in $[\delta, \infty)$ for any $\delta > 0$ such that $x$ solves (1.24), with $x(0) = X$ for almost every $t > 0$. (The right differentiability in Theorem 1.16 holds for $t > 0$.) See [18] for further details.

**(ii)** For $x_0 \in H$, let $x_\tau$ be the solution of $x + \tau \partial E(x) \ni x_0$. This is a unique minimizer of $E_\tau$, as shown in the proof of Lemma 1.19. Thus,

$$\sup_{\tau > 0} E_\tau(x_\tau) \leq E_\tau(x_0) = E(x_0) < \infty.$$

If $x_0 \in D(E)$, this bound implies $x_\tau \to x_0$ in $H$. Since $x_\tau \in D(\partial E)$, this implies that $D(E) \subset \overline{D(\partial E)}$. By definition, $D(E) \supset D(\partial E)$. Thus, $\overline{D(E)} = \overline{D(\partial E)}$. Such a simple proof is given in [6, Chapter 2, p. 48].

## 1.2.4   Gradient Flow of a Convex Function-Energetic Variational Inequality

The notion of the gradient flow of a convex function can be extended in a complete metric space, which is more general than a Hilbert space. We conclude Sect. 1.2 by defining an evolutionary variational inequality, which is an extended notion of $\dot{x} \in -\partial E(x)$, and by proving the uniqueness of a solution.

We consider the gradient flow $\dot{x} \in -\partial E(x)$, i.e.,

$$E(x(t) + h) - E(x(t)) \geq \langle h, -\dot{x}(t) \rangle$$

for all $h \in H$. We set $h = z - x(t)$ and observe that

$$\frac{1}{2} \frac{d}{dt} \|x(t) - z\|^2 = \langle x(t) - z, \dot{x}(t) \rangle = \langle z - x(t), -\dot{x} \rangle$$

$$\leq E(z) - E(x(t)), \quad z \in H.$$

This leads the definition of the gradient flow $\dot{x} \in -\partial E(x)$ in a general complete metric space $M$. We say a curve $x : (a, b) \to M$ satisfies the *evolutionary variational inequality* for $\dot{x} \in -\partial E(x)$ if

$$\frac{1}{2} \frac{d}{dt} \operatorname{dist}(x(t), z)^2 \leq E(z) - E(x(t)) \quad \text{for all} \quad z \in M, \tag{1.26}$$

where dist denotes the distance in $M$. We may weaken this definition by considering its integral form

$$\frac{1}{2}\left(\text{dist}\,(x(t), z)^2 - \text{dist}\,(x(s), z)^2\right) \le \int_s^t (E(z) - E\,(x(\tau)))\,d\tau, \quad z \in M,$$

where $t, s \in (a, b)$, $t \ge s$. The point is that a solution satisfying (1.26) is unique. Indeed, let $x_i = x(t, X_i)$ $(i = 1, 2)$ be a solution satisfying (1.26) starting from $X_i$. Then

$$\frac{1}{2}\frac{d}{dt}\,\text{dist}\,(x_1(t), x_2(t))^2$$

$$= \frac{1}{2}\frac{d}{dt}\,\text{dist}\,(x_1(t), x_2(s))^2\Big|_{s=t} + \frac{1}{2}\frac{d}{dt}\,\text{dist}\,(x_1(s), x_2(t))^2\Big|_{s=t}$$

$$\le E\,(x_2(t)) - E\,(x_1(t)) + E\,(x_1(t)) - E\,(x_2(t)) = 0.$$

This implies that

$$\text{dist}\,(x_1(t), x_2(t)) \le \text{dist}(X_1, X_2).$$

In particular, the solution $x$ satisfying (1.26) with $x|_{t=0} = X$ is unique.

The notion of a solution of (1.26) can be extended to semiconvex functional $E$. Instead of (1.26), for $\lambda$-convex $E$, we still say that $x$ is a solution of the *evolutionary variational inequality* for $\dot{x} \in -\partial E(x)$ if

$$\frac{1}{2}\frac{d}{dt}\,\text{dist}\,(x(t), z)^2 \le E(z) - E\,(x(t)) - \frac{\lambda}{2}\,\text{dist}\,(x(t), z)^2$$

for all $z \in X$. This solution is also unique because

$$\text{dist}\,(x_1(t), x_2(t))^2 \le \text{dist}(X_1\ X_2)^2 \exp(-\lambda t).$$

For a precise definition of an evolutionary variational inequality as well as $\lambda$-convexity, the reader is referred to the well-written book by [3]. There, the absolute continuity of the curve $x$ is assumed, so that dist $(x(t), z)$ is differentiable for almost all $t > 0$.

## 1.2.5 Simple Examples

The theory of gradient flow of a convex function has wide application to partial differential equations, and it gives a rigorous meaning to a solution. In this subsection, we present a few examples closely related to Sobolev spaces. For the basic properties of these spaces, the reader is referred to basic textbooks, for example, [19, Chapter 8, 9] and [36, Chapter 5]. To avoid discussing boundary conditions, we consider functionals defined on the space of periodic functions.

By a function $f$ on $\mathbf{T}^N = \prod_{i=1}^N (\mathbf{R}/\omega_i \mathbf{Z})$, with $\omega_i > 0$ ($i = 1, \ldots, N$), we mean that $f$ is a function on $\mathbf{R}^N$ satisfying

$$f(x + \omega_i e_i) = f(x) \quad \text{for all} \quad x \in \mathbf{R}^N,$$

where $\{e_i\}_{i=1}^N$ is the standard basis of $\mathbf{R}^N$. In other words, $f(x_1, \ldots, x_N)$ is periodic in $x_i$ with period $\omega_i$. Here, $\mathbf{Z}$ denotes the set of all integers. For $p \in [1, \infty]$, let $L^p(\mathbf{T}^N)$ denote the space of all real-valued $p$th *integrable* functions $f$ on $\mathbf{T}^N$ equipped with the *norm*

$$\|f\|_p = \left( \int_{\mathbf{T}^N} |f(x)|^p \, dx \right)^{1/p} = \left( \int_0^{\omega_1} \cdots \int_0^{\omega_N} |f(x)|^p \, dx \right)^{1/p}, \quad 1 \le p < \infty,$$

$$\|f\|_\infty = \operatorname*{ess.\,sup}_{x \in \mathbf{T}^N} |f(x)|, \quad p = \infty.$$

Here, $\alpha = \operatorname{ess.\,sup}_{x \in \mathbf{T}^N} |f(x)|$ (*essential supremum* of $f$ on $\mathbf{T}^N$) is defined by

$$\alpha = \inf \left\{ \beta \mid \mathcal{L}^N (\{|f| > \beta\}) \right\},$$

where $\mathcal{L}^N$ denotes the $N$-dimensional *Lebesgue measure* in $\mathbf{R}^N$ and $\{|f| > \beta\}$ denotes the set

$$\left\{ x = \left( x^1, \ldots, x^N \right) \mid 0 \le x^i < \omega_i, \ i = 1, \ldots, N, \ |f(x)| > \beta \right\}.$$

When we write $L^p(\mathbf{T}^N)$, we always identify functions whose values agree with each other *almost everywhere* (a.e.) so that $L^p(\mathbf{T}^N)$ is a normed space. It is well known that $L^p(\mathbf{T}^N)$ is a *Banach space*. The space $L^2(\mathbf{T}^N)$ is a Hilbert space equipped with the inner product

$$\langle f, g \rangle = \int_{\mathbf{T}^N} f(x)g(x)dx, \quad f, g \in L^2(\mathbf{T}^N),$$

so that $\|f\|_2 = \langle f, f \rangle^{1/2}$.

Let $W^{1,p}(\mathbf{T}^N)$ denote the $L^p$-*Sobolev space* of order 1. In other words,

$$W^{1,p}(\mathbf{T}^N) = \left\{ f \in L^p(\mathbf{T}^N) \mid D_j f \in L^p(\mathbf{T}^N), \ j = 1, \ldots, N \right\},$$

where $g_j = D_j f$ denotes the distributional derivative of $f$ in $x_j$, i.e.,

$$\int_{\mathbf{T}^N} f \frac{\partial \varphi}{\partial x_j} dx = - \int_{\mathbf{T}^N} g_j \varphi dx$$

for all $\varphi \in C_c^\infty(\mathbf{T}^N)$, where $C_c^\infty(\Omega)$ denotes the space of all smooth functions in $\Omega$ with compact support in $\Omega$. Since $\mathbf{T}^N$ is compact, $C_c^\infty(\mathbf{T}^N)$ agrees with $C^\infty(\mathbf{T}^N)$, the space of all smooth functions in $\mathbf{T}^N$. It is well known that $W^{1,p}(\mathbf{T}^N)$ is a Banach space equipped with the norm

$$\|f\|_{W^{1,p}} = \left(\|f\|_p^p + \sum_{j=1}^N \|D_j f\|_p^p\right)^{1/p}, \quad 1 \le p < \infty,$$

$$\|f\|_{W^{1,\infty}} = \max_{1\le j\le N}\left(\|f\|_\infty, \|D_j f\|_\infty\right), \quad p = \infty.$$

The space $W^{1,\infty}(\mathbf{T}^N)$ can be identified with $\mathrm{Lip}(\mathbf{T}^N)$, the space of all Lipschitz continuous functions equipped with the norm $\|f\|_\infty + [f]_{\mathrm{Lip}}$; see, for example, [36, §5.8. b].

We next consider a couple of functionals.

**A. $p$-Dirichlet energy for $1 < p < \infty$**

As a Hilbert space, we set $H = L^2(\mathbf{T}^N)$. Let $E_p(u)$, $u \in H$ denote the $p$-Dirichlet energy defined by

$$E_p(u) = \begin{cases} \dfrac{1}{p}\displaystyle\int_{\mathbf{T}^N} |\nabla u|^p \mathrm{d}x, & u \in W^{1,p}(\mathbf{T}^N) \cap H, \\ \infty, & \text{otherwise.} \end{cases}$$

Here, $\nabla u$ denotes the distributional gradient $\nabla u = (D_1 u, \ldots, D_N u)$ and $|\nabla u|^2 = \sum_{j=1}^N |D_j u|^2$.

---

**Lemma 1.21**

*Assume that $1 < p < \infty$. The functional $E_p$ is lower semicontinuous in $H$ and convex.*

---

*Proof.* The convexity is easy to prove since $|y|^p$ is a convex function. It remains to prove the lower semicontinuity

$$E_p(u) \le \liminf_{k\to\infty} E_p(u_k) \text{ for } u_k \to u \text{ (as } k \to \infty) \text{ in } H. \tag{1.27}$$

We may assume that the right-hand side is finite so that $u_k \in W^{1,p}(\mathbf{T}^N)$ and $\sup_{k\ge 1} E_p(u_k)$ is finite by taking a suitable subsequence. We first prove that $E_p(u)$ is finite. It suffices to prove that $u \in W^{1,p}(\mathbf{T}^N)$ if $\sup_{k\ge 1}\|D_j u_k\|_p < \infty$ for all $j = 1, 2, \ldots, N$. By *-weak compactness, a bounded sequence $\{f_k\}$ in $L^p(\mathbf{T}^N)$ has a $*$-weak convergent subsequence $\left\{ f_{k_m} \right\}$ (Exercise 1.9, [19, Theorem 3.16], [90]). Since $L^p(\mathbf{T}^N)$ is the *dual space* of $L^{p'}(\mathbf{T}^N)$ with $1/p + 1/p' = 1$ for $1 < p \leq \infty$ (see [19, Section 4.3]), this means that there is some $f \in L^p(\mathbf{T}^N)$ such that

$$\lim_{m \to \infty} \int_{\mathbf{T}^N} f_{k_m} \varphi \, dx = \int_{\mathbf{T}^N} f \varphi \, dx$$

for all $\varphi \in L^{p'}(\mathbf{T}^N)$. Since $\|D_j u_k\|_p$ is bounded, we apply this property to the sequence $\{D_j u_k\}$ to get a subsequence $\left\{ D_j u_{k_m} \right\}$ and $g_j \in L^p(\mathbf{T}^N)$ such that

$$\lim_{m \to \infty} \int_{\mathbf{T}^N} D_j u_{k_m} \varphi \, dx = \int_{\mathbf{T}^N} g_j \varphi \, dx.$$

We take $\varphi \in C_c^\infty(\mathbf{T}^N) \subset L^{p'}(\mathbf{T}^N)$ and observe that

$$\lim_{m \to \infty} \left( - \int_{\mathbf{T}^N} u_{k_m} \frac{\partial \varphi}{\partial x_j} \, dx \right) = - \int_{\mathbf{T}^N} u \frac{\partial \varphi}{\partial x_j} \, dx$$

since $u_k \to u$ in $H = L^2(\mathbf{T}^N)$. By the definition of $D_j u_{k_m}$,

$$- \int_{\mathbf{T}^N} u_{k_m} \frac{\partial \varphi}{\partial x_j} \, dx = \int_{\mathbf{T}^N} \left( D_j u_{k_m} \right) \varphi \, dx.$$

Sending $k \to \infty$, we now observe that

$$- \int_{\mathbf{T}^N} u \frac{\partial \varphi}{\partial x_j} \, dx = \int_{\mathbf{T}^N} g_j \varphi \, dx$$

for all $\varphi \in C_c^\infty(\mathbf{T}^N)$. In other words, $D_j u = g_j \in L^p(\mathbf{T}^N)$, so $g_j$ is independent of the choice of a subsequence $\{u_{m_k}\}$. (Thus, $D_j u_k$ converges $*$-*weakly* to $D_j u$ without taking a subsequence.) It remains to prove that $u \in L^p(\mathbf{T}^N)$. For $1 < p \leq 2$, this is trivial since $L^p(\mathbf{T}^N) \subset H = L^2(\mathbf{T}^N)$. We first recall the *Poincaré inequality* [36, §5.8.1],

$$\|u_k - (u_k)_{av}\|_p \leq C \left( E_p(u_k) \right)^{1/p},$$

where $(u_k)_{av}$ is the average of $u_k$ over $\mathbf{T}^N$, i.e.,

$$(u_k)_{av} = \frac{1}{|\mathbf{T}^N|} \int_{\mathbf{T}^N} u_k \, dx, \quad |\mathbf{T}^N| = \omega_1 \cdots \omega_N.$$

Here, $C$ is a constant depending only on $p$ and $|\mathbf{T}^N|$. (The Poincaré inequality is valid for $1 \leq p < \infty$, but we keep $1 < p < \infty$ in the following argument.) By the

*Hölder inequality* (which is called the Schwarz inequality in this case), we get

$$|(u)_{\text{av}}| \leq |\mathbf{T}^N|^{-1}\|1\|_2\|u\|_2 = |\mathbf{T}^N|^{-1/2}\|u\|_2.$$

Thus, applying the Poincaré inequality yields

$$\|u_k\|_p \leq \|u_k - (u_k)_{\text{av}}\|_p + \|(u_k)_{\text{av}}\|_p$$

$$\leq C\left(E_p(u_k)\right)^{1/p} + |\mathbf{T}^N|^{-\frac{1}{2}+\frac{1}{p}}\|u_k\|_2.$$

Since $u_k \to u$ in $L^2(\mathbf{T}^N)$ as $k \to \infty$ so that $\|u_k\|_2$ is bounded, we now conclude that $\sup_{k\geq 1}\|u_k\|_p < \infty$. By $*$-weak compactness as previously, there is a subsequence $\{u_{k_m}\}$ of $\{u_k\}$ and $w \in L^p(\mathbf{T}^N)$ such that

$$\lim_{m\to\infty} \int_{\mathbf{T}^N} u_{k_m}\varphi \, dx = \int_{\mathbf{T}^N} w\varphi \, dx$$

for $\varphi \in L^{p'}(\mathbf{T}^N)$. For $\varphi \in C_c^\infty(\mathbf{T}^N)$,

$$\lim_{k\to\infty} \int_{\mathbf{T}^N} u_k\varphi \, dx = \int_{\mathbf{T}^N} u\varphi \, dx$$

since $u_k \to u$ in $L^2(\mathbf{T}^N)$ as $k \to \infty$. Thus,

$$\int_{\mathbf{T}^N} (w - u)\varphi \, dx = 0$$

for all $\varphi \in C_c^\infty(\mathbf{T}^N)$. By the fundamental lemma of the calculus of variations (see Exercise 2.3 or [19, Corollary 4.24]), this implies $w = u$ almost everywhere (a.e.); in particular, $u \in L^p(\mathbf{T}^N)$. We thus conclude that $u \in W^{1,p}(\mathbf{T}^N)$.

By the lower semicontinuity of $L^p$-norm under $*$-*weak convergence* ( [19, Proposition 3.13], [90]), we conclude (1.27).                                    □

We thus apply Theorem 1.14 with $H = L^2(\mathbf{T}^N)$ to solve $u_t \in -\partial E_p(u)$ for $1 < p < \infty$. In the case $p = 2$, the case of the Dirichlet energy is simply the heat equation. For general $p \in (1,\infty)$, the equation $u_t \in -\partial E_p(u)$ is simply the $p$-Laplace diffusion equation $u_t = \text{div}\left(|\nabla u|^{p-2}\nabla u\right)$. This can be seen easily at least formally as follows. We see

$$\lim_{\varepsilon\downarrow 0}\left(E_p(u + \varepsilon h) - E_p(u)\right)/\varepsilon = \int_{\mathbf{T}^N} |\nabla u|^{p-2}\nabla u \cdot \nabla h dx$$

$$= -\int_{\mathbf{T}^N}\left(\text{div}\left(|\nabla u|^{p-2}\nabla u\right)\right)h dx$$

by integration by parts.

## B. Total variation energy

We next consider the total variation energy on $H = L^2(\mathbf{T}^N)$:

$$E(u) = \begin{cases} \int_{\mathbf{T}^N} |\nabla u| dx, & D_j u \in L^1(\mathbf{T}^N) \ (j = 1, \dots, N), \\ \infty, & \text{otherwise.} \end{cases}$$

We note that this $E$ is convex but not lower semicontinuous even for one-dimensional problems. In fact, it suffices to construct a special sequence. We may assume that $N = 1$ and $\omega_1 = 1$. We consider the sequence

$$u_k(x) = \begin{cases} 1, & |x| < \dfrac{1}{4}, \\ k\left(\dfrac{1}{4} + \dfrac{1}{k} - |x|\right)_+, & \dfrac{1}{4} \le |x| \le \dfrac{1}{2}. \end{cases}$$

This sequence converges to

$$u_0(x) = \begin{cases} 1, & |x| < \dfrac{1}{4}, \\ 0, & \dfrac{1}{4} \le |x| \le \dfrac{1}{2}, \end{cases}$$

in $L^2(\mathbf{T}^1)$ and $E(u_k) = 2$; see Fig. 1.7. However, the distributional derivative $D_1 u$ does not belong to $L^1(\mathbf{T}^N)$, i.e., $u_0 \notin W^{1,1}(\mathbf{T})$. In other words, $E(u_0) = \infty$. Thus, $E$ is not lower semicontinuous. For a better formulation, we introduce total variation for more general functions than $W^{1,1}(\mathbf{T}^N)$.

**Fig. 1.7** Graphs of $u_0$ and $u_k$ $(k = 1, 2, \dots)$

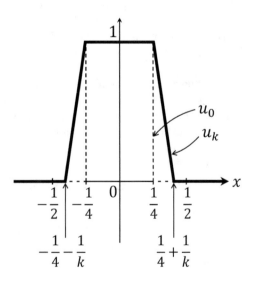

For $f \in L^1(\mathbf{T}^N)$, we define the (essential) *total variation* of $f$ by

$$\int_{\mathbf{T}^N} |\nabla f| := \sup \left\{ \int_{\mathbf{T}^N} f \operatorname{div} \varphi \mathrm{d}x \ \middle| \ \|\varphi\|_\infty \leq 1, \ \varphi \in \left( C_c^\infty(\mathbf{T}^N) \right)^N \right\},$$

where $\varphi = (\varphi_1, \ldots, \varphi_N)$. The totality of $f \in L^1(\mathbf{T}^N)$ having finite total variation is denoted by $BV(\mathbf{T}^N)$. If one interprets $E$ by

$$E_1(u) = \begin{cases} \displaystyle\int_{\mathbf{T}^N} |\nabla u|, & u \in BV(\mathbf{T}^N) \cap H \\ \infty, & \text{otherwise,} \end{cases}$$

then this $E_1$ is a lower semicontinuous convex function. This gives a rigorous formulation of the total variation flow $u_t = \operatorname{div}(\nabla u / |\nabla u|)$ by $u_t \in -\partial E_1(u)$. The total variation equation is quite different from the $p$-Laplace diffusion equation. The reader is referred to the textbook [5] for an analysis of the total variation equation. Our unique existence result (Theorem 1.14), together with Remark 1.20, guarantees the unique solvability for $u_t \in -\partial E_p(u)$ $(1 \leq p < \infty)$ for any initial data $u_0 \in L^2(\mathbf{T}^N)$ since $D(E_p)$ is dense in $H$; see, for example, [36, §5.3] for the density.

## 1.3 Notes and Comments

### 1.3.1 More on Uniqueness on Continuous Vector Field $b$

Condition (1.10) was introduced by Osgood [80] in the nineteenth century. As noted in Remark 1.4, one can weaken the assumption in time in (1.10). Indeed, the right-hand side of (1.10) can be replaced by $M(t)f(|x - y|)$, where $M(t)$ is integrable on $I$. This type of uniqueness result is often called Motel–Tonelli's uniqueness theorem, which went back to Tonelli [85]; see, for example, [1, Theorem 1.5.1] or [52, Theorem 6.1]. However, in this situation, $M(t)$ is not allowed to be equal to $1/t$ because $\int_0^1 \mathrm{d}t/t = \infty$. Nevertheless, there is a uniqueness criterion called Nagumo's theorem [77], which asserts that uniqueness follows if one assumes $[b(\cdot, t)]_{\mathrm{Lip}(\mathbf{R}^N)} \leq 1/t$, $t \in \mathring{I}$; note that $1/t$ cannot be replaced by $M/t$. There are several generalizations of Nagumo's theorem and Motel–Tonelli's theorem. The reader is referred to [24], [23], [25] for a more recent development of this topic. In particular, in [25], a convex combination of Nagumo's condition and Osgood's condition is considered.

So far we have discussed several sufficient conditions to guarantee uniqueness. There is an interesting necessary and sufficient condition for uniqueness discovered by Okamura [79].

**Theorem 1.22**
*Assume that $b = b(x, t)$ is continuous in $B_R \times \overline{I}$, with $I = [0, a)$, $a > 0$. For any $X \in \overset{\circ}{B}_R$ there is at most one solution of (1.1) with initial datum $X$ if and only if there exists a nonnegative function $V \in C^1(U)$ satisfying the following conditions:*

**(i)** *For a given point $(x, y, t) \in U$, $V(x, y, t) = 0$ is equivalent to $x = y$;*
**(ii)** $\dfrac{\partial V}{\partial t} + b(x, t) \cdot \nabla_x V + b(y, t) \cdot \nabla_y V \le 0$ *in $U$, where $U = \overset{\circ}{B}_R \times \overset{\circ}{B}_R \times \overset{\circ}{I}$.*

Here, $\nabla_x$ (resp. $\nabla_y$) denotes the gradient in variable $x$ (resp. $y$), i.e., $\nabla_x = (\partial/\partial x_1, \ldots, \partial/\partial x_N)$ (resp. $\nabla_y = (\partial/\partial y_1, \ldots, \partial/\partial y_N)$). The function $V = V(x, y, t)$ plays the role of a kind of Lyapunov function. The proof that the existence of $V$ implies the uniqueness is easy. Indeed, let $x_i = x_i(t)$ $(i = 1, 2)$ be a solution of (1.1) with initial datum $X_i$. We set $\Phi(t) = V(x_1(t), x_2(t), t)$ for $t \in I$, and by (ii) we observe that

$$\frac{d\Phi}{dt}(t) = \frac{\partial V}{\partial t} + b(x, t) \cdot \nabla_x V + b(y, t) \cdot \nabla_y V \Big|_{(x, y, t) = (x_1(t), x_2(t), t)} \le 0, \quad t \in \overset{\circ}{I}.$$

Since $\Phi(0) = V(X, X, 0) = 0$ by (i), this implies $\Phi(t) \le 0$ for $t \in I$. However, since $\Phi \ge 0$, this implies that $\Phi \equiv 0$. By (i), this implies $x_1(t) = x_2(t)$ for $t \in I$.

The converse is more involved. Here, we only give an idea of a proof. We introduce a distance-like function $d(P, Q)$ defined for $P = (x_0, y_0, t_0)$, $Q = (x_*, y_*, t_*)$, with $t_0 < t_*$, where $P, Q \in U$. We divide the interval $[t_0, t_*]$ such that $t_0 < t_1 < \ldots < t_m = t_*$. Let $\Delta = \{t_0 < t_1 < \cdots < t_m\}$ denote its division. On each interval $[t_{k-1}, t_k]$ we consider a solution curve $(x, y) = (x(t), y(t))$ of

$$\frac{dx}{dt} = b(x, t), \quad \frac{dy}{dt} = b(y, t).$$

We denote the left endpoint by $P_k = (x(t_{k-1} + 0), y(t_{k-1} + 0), t_{k-1}) \in U$ and the right endpoint by $Q_k = (x(t_k - 0), y(t_k - 0), t_k) \in U$ (Fig. 1.8); here, $x(t \pm 0) = \lim_{\varepsilon \downarrow 0} x(t \pm \varepsilon)$. We then consider the sum of jumps, i.e.,

$$S_\Delta = \sum_{j=1}^{m+1} |P_j - Q_{j-1}|, \quad Q_0 = P, \quad P_{m+1} = Q.$$

Note that the value $S_\Delta$ may not be unique even if $\Delta$ is fixed. For given $P, Q \in U$, there is at least one division $\Delta$ and $P_1, \ldots, P_m, Q_1, \ldots, Q_m$ such that $P_k$ and $Q_k$ are connected by a solution curve on $[t_{k-1}, t_k]$. Indeed, there always exists a local-in-time solution to (1.1) when $b$ is simply continuous, which is known as Peano's

**Fig. 1.8** Location of $P_k$ and $Q_k$

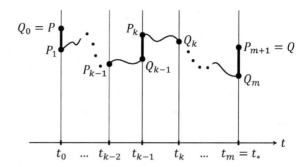

existence theorem; for the proof, see [52, Chapter 2]. Checking an existence time estimate, we conclude that if all subintervals are sufficiently small uniformly and $P_k = (0, 0, t_{k-1})$ $(1 \le k \le m)$, then there is a solution curve $(x, y) = (x(t), y(t))$ on $[t_{k-1}, t_k]$ starting from $P_k$. We denote by $Q_k$ the end point at $t_k$, i.e., $Q_k = (x(t_k), y(t_k), t_k)$. We set

$$d(P, Q) = \inf \left\{ S_\Delta \,\middle|\, \text{choice of } \Delta \text{ and} \right.$$

$$\left. \text{a solution curve } (x, y) = (x(t), y(t)) \text{ on each subinterval} \right\}.$$

For $t_* < t_0$ we set $d(P, Q) = d(Q, P)$. It turns out that this $d$ satisfies $d(P, Q) \le d(P, R) + d(R, Q)$ provided that the $t$-coordinate $t_P$ of $P$, $t_Q$ of $Q$, $t_R$ of $R$ fulfills $t_P < t_R < t_Q$. Moreover, $d(P, Q) = 0$ if and only if there is at least one solution curve connecting $P$ and $Q$ on $[t_0, t_*]$. In other words, there is a solution

$$\frac{dx}{dt} = b(x, t), \quad \frac{dy}{dt} = b(y, t) \quad \text{on} \quad [t_0, t_*],$$

with $P = (x(t_0), y(t_0), t_0)$ and $Q = (x(t_*), y(t_*), t_*)$. Furthermore, the value $d(P, Q)$ does not change even if $Q$ is replaced by a point on a solution curve connecting $Q$. We set

$$\tilde{V}(x, y, t) = \inf \left\{ d\left(P, (x, y, t)\right) \,\middle|\, P = (X, X, s), s < t, X \in \overset{\circ}{B}_R \right\}.$$

This is close to what we would like to obtain; this $\tilde{V}$ is not $C^1$, so one must mollify to obtain the desired $V$. Property (ii) follows from the nonincreasing property of $d$ along solution curves. Property (i) follows from the uniqueness of a solution.

This is an argument of doubling variables that plays an important role in showing the uniqueness of an entropy solution as well as a viscosity solution for first-order partial differential equations. We shall discuss these topics in Chaps. 3 and 4. It is also interesting to relate to ordinary differential equations through the idea of characteristics.

The Lyapunov-like function $V$ in Theorem 1.22 is also important to characterize the infinitesimal generator (defined later) of a Lipschitz semigroup. Let $W$ be a Banach space. We say that $\{S(t) \mid t \geq 0\}$ is a *Lipschitz semigroup* on $Z \subset W$ if $S(t)$ is a mapping from $Z$ to $Z$ satisfying the following conditions:

(a) $S(0)x = x$ $(x \in Z)$, $S(t+s)x = S(t)S(s)x, t \geq 0, x \in Z$;
(b) $S(t)x$ is continuous in time on $[0, \infty)$ with values in $W$ for each $x \in Z$;
(c) For each $t > 0$, there is $K$ such that

$$\|S(t)x - S(t)y\| \leq K\|x - y\|, \quad x, y \in Z, \quad t \in [0, t).$$

Its *infinitesimal generator* $b(x)$ is defined by

$$b(x) = \lim_{h \downarrow 0} (S(h)x - x)/h$$

provided that the strong limit exists in $W$. Let $b$ be a given continuous operator from a closed set $Z$ to $W$. In [61], a necessary and sufficient condition that $b$ is an infinitesimal generator of a Lipschitz semigroup is given by using a function like $V$ in Theorem 1.22 with the semitangent condition of $Z$. In other words, their condition is a condition so that Eq. (1.1), with $b$ independent of $t$, is in some sense uniquely solvable in a Banach space. This result is extended to the case where $b$ depends also on time, i.e., nonautonomous case [62].

We conclude this subsection by mentioning the uniqueness results for equations with fractional time derivatives. The uniqueness results are now available under the Osgood condition, even for a single higher-order equation; see the recent paper [82] for $\alpha > 1$ and papers cited therein. However, compared with usual ordinary differential equations, our understanding remains limited. For example, Montel–Tonelli type results are not yet available.

### 1.3.2 Forward Uniqueness on Discontinuous Vector Field $b$

If we only consider forward uniqueness, our $b$ is allowed to be discontinuous. The unique solvability result (Theorem 1.14) goes back to Kōmura [65] and was well developed by Brezis and others; see [6, 18], and [66]; see also [36, §9.6]. These books are general references for Sect. 1.2. From the point of forward uniqueness, the vector field $b$ in (1.1) is allowed to have some discontinuity in $x$ and $t$. Here is a condition slightly weaker than monotonicity to guarantee uniqueness. Instead of (1.24), we consider the problem

$$\dot{x} \in -A(t)x, \tag{1.28}$$

where $A(t)$ is a time-dependent set-valued operator in a Hilbert space $H$.

**Theorem 1.23**

*Assume that there is a function* $k = k(t)$ *which is integrable on* $I = [0, a)$ *satisfying*

$$\langle x' - y', x - y \rangle + k(t)\|x - y\|^2 \geq 0 \qquad (1.29)$$

*for all* $x' \in A(t)x$, $y' \in A(t)y$, $x, y \in D(A(t))$, $t \in I$. *Then a solution* $x \in C(I, H)$ *(locally absolutely continuous in* $I$ *with values in* $H$*) of* (1.28) *with initial data* $X \in H$ *is unique.*

The proof is very similar to that of Theorem 1.15. We set $x_i = x(t, X_i)$ with $X_i \in D(A(0))$ for $i = 1, 2$. We argue similarly to get

$$\frac{1}{2}\frac{d}{dt}\|x_1 - x_2\|^2 \leq k(t)\|x_1 - x_2\|^2$$

for two solutions. By the Gronwall-type inequality (see Remark 2.10) this yields

$$\|x_1 - x_2\|^2(t) \leq \|X_1 - X_2\|^2 \exp\left(2\int_I k(t)dt\right).$$

This implies uniqueness. Condition (1.29) is often called the one-sided Lipschitz condition. For a general discontinuous right-hand side of (1.28), the reader is referred to the book by Deimling [31] and a more classical book by Filippov [40]. The latter already includes the one-sided Lipschitz condition when $H$ is of finite dimension.

The uniqueness issue remains of current interest. The reader is referred to the recent paper [13], where $b$ is discontinuous in $x$ (and also in $t$) across some interfaces.

### 1.3.3 Estimates of Lipschitz Constant

From Theorem 1.16 we expect that the solution $x$ in (1.24) has a Lipschitz bound:

$$\|x(t, X) - x(s, X)\| \leq \|A^0 X\|(t - s), \quad t > s \geq 0. \qquad (1.30)$$

Here, $A$ is as in Theorem 1.14. This is actually true. We give here an idea of the proof based on the construction of the solution. We admit the fact that the solution $x$ is constructed as a limit of the solution $x_\lambda$ of the approximate problem

$$\dot{x}_\lambda = A_\lambda x_\lambda, \quad x_\lambda(0) = X$$

as $\lambda \downarrow 0$, where $A_\lambda$ is the Yosida approximation of $A$; see Remark 1.18 (iii). Since $A_\lambda = AJ_\lambda$ is Lipschitz globally in time, this problem admits a globally-in-time solution with values in $H$. The solution satisfies a contraction principle by Theorem 1.15, i.e.,

$$\|x_\lambda(s, X) - x_\lambda(s, Y)\| \le \|X - Y\|, \quad t > s \ge 0,$$

since $A_\lambda$ is a monotone operator. Setting $Y = x_\lambda(t, X)$, $t > 0$ and using the semigroup property $x_\lambda(s, x_\lambda(\tau, X)) = x_\lambda(s + \tau, X)$ $(s, \tau \ge 0)$, we see that

$$\|x_\lambda(s + \tau, X) - x_\lambda(\tau, X)\| \le \|x_\lambda(\tau, X) - X\|. \tag{1.31}$$

Since $x_\lambda$ solves

$$\frac{x_\lambda(\tau) - X}{\tau} = \frac{1}{\tau} \int_0^\tau (A_\lambda x_\lambda)(\sigma) d\sigma,$$

sending $\tau \to 0$ implies

$$\liminf_{\tau \downarrow 0} \left\| \frac{x_\lambda(\tau) - X}{\tau} \right\| \le \|A_\lambda X\| \le \|A^0 X\|;$$

the last inequality follows from (1.25). By (1.31), we see that

$$\|\dot{x}_\lambda\| \le \|A^0 X\|.$$

Thus,

$$\|x_\lambda(t, X) - x_\lambda(s, X)\| \le \|A^0 X\|(t - s), \quad t, s > 0. \tag{1.32}$$

If we admit that $x$ is given as a limit of $x_\lambda$ (as $\lambda \downarrow 0$) in $C([0, T], H)$, $T \in (0, \infty)$, i.e.,

$$\lim_{\lambda \downarrow 0} \sup_{0 < t < T} \|x(t) - x_\lambda(t)\| = 0,$$

then (1.32) implies (1.30).

### 1.3.4   A Few Directions for Applications

It is very popular to use ordinary differential equations (1.1) to describe time-dependent phenomena in various areas of science and technology. In a planar domain, a velocity field $b$ of incompressible fluids with bounded vorticity satisfies the Osgood condition but may not satisfy the Lipschitz condition. For the Euler equation describing the motion of incompressible inviscid fluids, a global-in-time (weak) unique solution was first constructed by Wolibner [89]. For this purpose, he solved (1.1) for a velocity field $b$ satisfying the Osgood condition. In his setting, the vorticity field is bounded and continuous. An elementary introduction to the Euler equations is found in [78, Chapter 7]. For the Biot–Savart law and the vorticity, the reader is referred to [45, Chapter 2]. In [45, Chapter 2], answers are given in what follows to Exercises 1.3 and 1.4. The theory of DiPerna–Lions in Chap. 2 is also introduced for the study of three-dimensional Euler equations.

Section 1.2 contains various applications to nonlinear partial differential equations, especially when the equation is quasi-linear, including the total variation flow equation. Total variation flow is used in image analysis; see [5].

The equations with fractional time derivatives discussed in Sect. 1.1.5 are fundamental to studying a diffusion equation with fractional time derivatives. Such an equation is useful for describing the anomalous diffusion found in contaminated water diffusing in soil. See the classic book on fractional differential equations by Podlubny [83]; see also the recent book by Kubica, Ryszewska, and Yamamoto [69] for the theory of diffusion equations with fractional time derivatives. It turns out that the equations with fractional time derivatives are actually derived as a kind of the Dirichlet-Neumann map of a diffusion equation as recently observed in [48].

## 1.4 Exercises

1.1 Assume that a function $b$ on $\mathbf{R}^N$ is $C^1$. Prove that $b$ is locally Lipschitz, i.e., $[b]_{\mathrm{Lip}(B_R)} < \infty$ for every $R > 0$.

1.2 Prove Remark 1.4. For simplicity, assume that a continuous function $b = b(x, t)$ satisfies

$$|b(x, t) - b(y, t)| \le M(t)|x - y|, \quad t \in (0, a),$$

for $x, y$ such that $|x|, |y| < R$, where $R$ is a fixed positive number. Assume that $M(t)$ is integrable on $(0, a)$. Prove that a solution $x$ of (1.1) with initial datum $X$, with $|X| < R$ in $(0, a)$, is unique provided $|x| < R$ in $(0, a)$.

1.3 Let $g$ be smooth in $\mathbf{R}^N$ (with $N = 2$ or 3) and compactly supported. Show that $E * g$ is smooth, where $E$ is the fundamental solution of $-\Delta$.

1.4 Under the same assumption made in Exercise 1.3, prove that $-\Delta(E * g) = g$.

1.5 Consider $\varphi = \varphi(x_1, x_2)$ such that

$$\varphi(x_1, x_2) = x_1 x_2 \left(-\log|x|\right)^\alpha, \quad x = (x_1, x_2) \in B_1 \subset \mathbf{R}^2,$$

with $\alpha \in (0, 1)$ in $B_1$, and $\varphi$ is smooth in $\mathbf{R}^2$ and compactly supported. Show that $u = \nabla^\perp \varphi$ satisfies $\operatorname{div} u = 0$. Show that $\operatorname{curl} u = -\Delta \varphi = \omega$ is continuous in $B_1$. The derivative should be interpreted in the sense of a Schwartz distribution, i.e.,

$$\int_{\mathbf{R}^2} (\operatorname{curl} u)\psi \, dx = \int_{\mathbf{R}^2} \omega \cdot \nabla^\perp \psi \, dx, \quad \psi \in C_c^\infty(\mathbf{R}^2)$$

for a scalar function. Show that $u$ is not Lipschitz continuous.

1.6 Assume that $A$ is a monotone operator in a Hilbert space $H$ and $\tau > 0$. Assume that $x_0 \in x + \tau Ax$ has a solution $x \in H$ for all $x_0 \in H$. Show that $A$ is a maximal monotone operator. (This is an easy part of Minty's theorem.)

1.7 Show that $Ax$ is a closed convex set if $A$ is a maximal monotone operator in a Hilbert space.

1.8 Set $E(x) = |x|$ on $H = \mathbf{R}$. Solve the initial value problem

$$\dot{x} \in -\partial E(x), \quad x(0) = X \in H.$$

1.9 Let $X$ be a separable Banach space and $X^*$ its dual space. We say that $x_m \in X^*$ converges to $x \in X^*$ *-weakly if

$$\langle x_m - x, y \rangle \to 0$$

for all $y \in X$ as $m \to \infty$. Here $\langle \, , \, \rangle$ denotes a canonical pair. An element $x^* \in X^*$ is a bounded linear functional $x^* : X \to \mathbf{R}$ and $\langle x^*, y \rangle := x^*(y)$ for $y \in X$. Show that if $\{x_m\}$ is bounded in $X^*$, then there is a *-weakly convergent subsequence of $\{x_m\}$. See, for example, [19, Theorem 3.16].

1.10 Let $E$ be a lower semicontinuous convex function in a Hilbert space $H$ with values in $\mathbf{R} \cup \{+\infty\}$. Assume that $E \not\equiv \infty$. For $\tau > 0$ and $x_0 \in H$, let $E_\tau$ be

$$E_\tau(x) = E(x) + \frac{1}{2\tau} \|x - x_0\|^2.$$

Then

$$\partial E_\tau(x) = \partial E(x) + \frac{1}{\tau}(x - x_0).$$

# Ordinary Differential Equations and Transport Equations

**2**

We continue to consider a system of ordinary differential equations (1.1), but we are more interested in the map $X \mapsto x(t, X)$, which is often called a *flow map* generated by a vector field $b$. If the initial value problem for (1.1) admits a unique local-in-time solution in a time interval $I = (0, a)$ with some $a > 0$ independent of $X$, the flow map is well defined. In Sect. 1.1, we gave a few sufficient conditions so that the flow map is uniquely determined assuming the existence of solutions to (1.1) with a given initial datum. Roughly speaking, if the vector field $b$ satisfies the Lipschitz condition or a weaker condition called the Osgood condition, then the flow map is well defined. Since the Lipschitz continuity of $b$ in $\mathbf{R}^N$ is equivalent to saying that the first distributional derivative of $b$ is in $L^\infty$ (see [36, §5.8. b]), it can be written $b \in W^{1,\infty}(\mathbf{R}^N)$, where the $W^{m,p}(\Omega)$ denotes the Sobolev space of order $m = 0, 1, 2, \ldots$ in $L^p(\Omega)$.

In this chapter, we are interested in the question of whether (1.2) can be replaced by $\|b(\cdot, t)\|_{W^{1,p}} \leq M$, with finite $p \geq 1$. However, unfortunately, this does not guarantee uniqueness; this can be easily seen if one elaborates Example 1.5. This suggests that we need some extra conditions for $b$ so that a flow map is well defined. It turns out that if div $b = 0$ or at least div $b$ is bounded, this is the case (under some growth assumptions on $b$ at the space infinity) provided that we regard the flow map $X \mapsto x(t, X)$ for almost all $X$ (almost everywhere (a.e.) $X$) in $\mathbf{R}^N$ not for all $X$. Such a theory was started by DiPerna and Lions [32] in the late 1980s.

In this section, we explain the uniqueness part of the theory of autonomous equations, i.e., $b$ is independent of time. To simplify the problem, we further assume that $b$ is periodic in space variables.

M.-H. Giga, Y. Giga, *A Basic Guide to Uniqueness Problems for Evolutionary Differential Equations*, Compact Textbooks in Mathematics, https://doi.org/10.1007/978-3-031-34796-2_2

## 2.1    Uniqueness of Flow Map

We consider a vector field (or $\mathbf{R}^N$-valued function) $b$ on $\mathbf{T}^N = \prod_{i=1}^{N} (\mathbf{R}/\omega_i \mathbf{Z})$, with $\omega_i > 0$ $(i = 1, \ldots, N)$, i.e.,

$$b(x) = \left( b^1(x), \ldots, b^N(x) \right) \quad \text{for} \quad x \in \mathbf{T}^N.$$

In other words, we assume $b$ is periodic in the $i$th variable with the period $\omega_i$. In this section, we always assume that

$$b^j \in W^{1,1}(\mathbf{T}^N) \quad \text{for} \quad 1 \le j \le N \quad \text{and} \quad \operatorname{div} b = 0 \quad \text{in} \quad \mathbf{T}^N. \tag{2.1}$$

We simply write the first condition by $b \in W^{1,1}(\mathbf{T}^N)$ instead of writing $b \in \left( W^{1,1}(\mathbf{T}^N) \right)^N$, though the latter is more precise notation. Here, $W^{1,p}(\mathbf{T}^N)$ is the $L^p$-Sobolev space introduced in Sect. 1.2.5. We are interested in discussing the uniqueness of a solution to (1.1) with $b$ independent of $t$, i.e.,

$$\dot{x} = b(x)$$

or

$$\frac{dx^i}{dt}(t) = b^i \left( x^1(t), \ldots, x^N(t) \right)$$

for $x(t) = \left( x^1(t), \ldots, x^N(t) \right)$ under condition (2.1). However, under condition (2.1), a flow map $X \mapsto x(t, X)$ (generated by $b$) for a fixed time $t$ may not be integrable on $\mathbf{T}^N$. In other words, each component of this map may not belong to $L^1(\mathbf{T}^N)$. To overcome this difficulty, we introduce a space

$$\mathcal{M} = \mathcal{M}(\mathbf{T}^N) := \left\{ \phi : \mathbf{T}^N \to \mathbf{R} \mid \text{(Lebesgue)} \; measurable \; \text{and} \right.$$

$$\left. |\phi| < \infty \; \text{a.e.} \right\}.$$

This space is metrizable. For example, if we define a *metric $d$* as

$$d(\phi, \psi) = \|\min (|\phi - \psi|, 1)\|_{L^1(\mathbf{T}^N)} \quad \text{for} \quad \phi, \psi \in \mathcal{M},$$

then $(\mathcal{M}, d)$ becomes a *metric space*. See Exercise 2.2 and 2.6. From this point forward, $\| \cdot \|_{L^p(\mathbf{T}^N)}$ (or $\| \cdot \|_{L^p}$) denotes the $L^p$-norm in $L^p(\mathbf{T}^N)$. The convergence in this metric corresponds to the *convergence in measure*, i.e., $d(\phi_j, \psi) \to 0$ as $j \to \infty$ implies for any $\delta > 0$

$$\mathcal{L}^N \left\{ x \in \mathbf{T}^N \mid |\phi_j - \psi|(x) > \delta \right\} \to 0 \quad \text{as} \quad j \to \infty,$$

where $\mathcal{L}^N$ denotes the $N$-dimensional Lebesgue measure; see Sect. 1.2.5 or Appendix 5.2 for a precise definition of a set in $\mathbf{T}^N$. For fixed $t$, we expect each component $x^i$ of a solution $x = x(t, X)$ belongs to $\mathcal{M}$ as a function of $X$, i.e., the mapping

$$x[t] : X \mapsto x(t, X)$$

is expected to be in $\mathcal{M}^N$. We also expect the mapping $x : t \mapsto x[t]$ to be defined for all $t \in \mathbf{R}$, and it is continuous from $\mathbf{R}$ to $\mathcal{M}^N$, i.e., $x \in C(\mathbf{R}, \mathcal{M}^N)$. The reason we expect $x$ to be defined for all $t$ is that the value $x[t](X) = x(t, X)$ actually belongs to the compact space $\mathbf{T}^N$, which prevents what is called blow-up phenomena.

If $b$ is divergence-free, i.e., solenoidal, then the flow map $x[t]$ must satisfy the volume-preserving property. In other words, for all $t \in \mathbf{R}$,

$$\mathcal{L}^N \left( \left\{ z \in \mathbf{T}^N \mid x[t]z \in A \right\} \right) = \mathcal{L}^N(A) \tag{2.2}$$

for any (Lebesgue) measurable set $A$. More generally,

$$\int_{\mathbf{T}^N} \psi\left(x(t, X)\right) \, \mathrm{d}X = \int_{\mathbf{T}^N} \psi(y) \, \mathrm{d}y$$

for any measurable function $\psi$ on $\mathbf{T}^N$. See Exercise 2.8. (In general, for a Lebesgue measurable set $A$, $f^{-1}(A) = \left\{ z \in \mathbf{T}^N \mid f(z) \in A \right\}$ may not be Lebesgue measurable for a Lebesgue measurable function $f$. The volume-preserving property implicitly guarantees that $x[t]^{-1}(A)$ will be Lebesgue measurable if $A$ is Lebesgue measurable.) The property (2.2) is obtained by div $b = 0$. Here is a formal argument assuming that $x$ is $C^1$ in $t$ and $X$. We set $F = (F_{ij}) = (\partial x^i / \partial X^j)$ for the flow map $x = x(t, X)$. (This is a *Jacobi matrix* of the flow map $X \mapsto x(t, X)$.) By the area formula (or change of variable of integration), to see (2.2), it suffices to prove that $\det F = 1$ for all $t$, where $\det F$ denotes the determinant of $F$. Let $\operatorname{tr} M$ denote the *trace* of $N \times N$ metrics $M$, i.e., it is the sum of the diagonal components of $M$. By elementary calculus, we see that

$$\frac{\mathrm{d}}{\mathrm{d}t} \det F = \det F \operatorname{tr} \left( \frac{\partial F}{\partial t} F^{-1} \right).$$

By Eq. (1.1), we see that

$$\frac{\partial F_{ij}}{\partial t} = \sum_{\ell=1}^{N} (\partial_\ell b_i)(x) F_{\ell j}.$$

Thus, $\dfrac{\mathrm{d}}{\mathrm{d}t} \det F = \operatorname{tr}(Db) \det F$. Here $Db$ denotes the Jacobian matrix $(Db)_{ij} = \partial_j b_i$, $1 \leq i, j \leq N$. We note that $\operatorname{tr}(Db) = \operatorname{div} b$. If div $b = 0$ so that $\operatorname{tr}(Db) = 0$, we now observe that $\det F$ is time independent. Since $\det F = 1$ at $t = 0$, we now conclude that $\det F = 1$ for all $t$. This formal argument is justified when the flow map $x[t] : X \mapsto x(t, X)$ is in $C^1(\mathbf{T}^N, \mathbf{T}^N)$. This is indeed true if $b$ is $C^1$, and it is known as $C^1$ dependence with respect to the initial data; See, for example, [52, Chapter 5].

We must consider a solution $x = x(t, X)$ of the ordinary differential equation (1.1), which is only continuous but may not be in $C^1$ in the time variable $t$. If one weakens the notion of a solution, there is a chance we lose the uniqueness. To keep the uniqueness, we consider a special class of a solution that is often called a renormalized solution. We consider a mapping $t \mapsto x(t, \cdot)$ from $\mathbf{R}$ to $\left(\mathcal{M}(\mathbf{T}^N)\right)^N$. If this mapping is continuous, we simply write $x \in C\left(\mathbf{R}, \left(\mathcal{M}(\mathbf{T}^N)\right)^N\right)$. It is also possible to consider the mapping $X \mapsto x(\cdot, X)$ from $\mathbf{T}^N$ to $(C(\mathbf{R}))^N$. This mapping is often called a flow map.

---

**Definition 2.1**

Assume that $x \in C\left(I, \left(\mathcal{M}(\mathbf{T}^N)\right)^N\right)$. We say that $x$ is a *(renormalized) solution* of (1.1) in $\mathbf{R}$ if

$$\frac{\partial}{\partial t}(\beta \circ x)(t, X) = D\beta\left(x(t, X)\right) b\left(x(t, X)\right) \qquad \text{on} \quad \mathbf{R} \times \mathbf{T}^N, \tag{2.3}$$

$$(\beta \circ x)|_{t=0}(X) = \beta(X) \qquad \text{on} \quad \mathbf{T}^N \tag{2.4}$$

for all $\beta \in C^1(\mathbf{T}^N, \mathbf{T}^N)$ such that $\beta \circ x \in L^\infty\left(\mathbf{R}, \left(\mathcal{M}(\mathbf{T}^N)\right)^N\right)$, where $\beta \circ x$ is a composite function defined by $(\beta \circ x)(t, X) = \beta\left(x(t, X)\right)$. Here $D\beta$ denotes the Jacobian matrix $(D\beta)_{ij} = \partial \beta^i / \partial x_j$, $1 \le i, j \le N$.

---

The time variable in (2.3) should be interpreted in the sense of a distribution whose variables are $t$ and $X$. In other words, (2.3) means that

$$-\int_{\mathbf{T}^N}\int_{-\infty}^{\infty} \frac{\partial \varphi}{\partial t}(t, X)(\beta \circ x)(t, X) \, dt \, dX$$

$$= \int_{\mathbf{T}^N}\int_{-\infty}^{\infty} \varphi(t, X) D\beta\left(x(t, X)\right) b\left(x(t, X)\right) \, dt \, dX$$

for all $\varphi \in C_c^\infty(\mathbf{T} \times \mathbf{T}^N)$. Of course, if $x$ is $C^1$ in $t$, then $x$ must satisfy (2.3) for all $\beta$ and $X$ if and only if $x$ is a solution to (1.1) with $x(0, X) = X$.

We need to explain the space $L^\infty\left(\mathbf{R}, \left(\mathcal{M}(\mathbf{T}^N)\right)^N\right)$. If $V$ is a Banach space $V$, then let $L^p(\mathbf{R}, V)$ be the space of all $p$th integrable functions on $\mathbf{R}$ as defined in Appendix 5.2 (4) using a Bochner integral. Since $\mathcal{M}(\mathbf{T}^N)$ is not a normed space but just a metric space, we must extend the definition. The space $L^\infty\left(\mathbf{R}, \mathcal{M}(\mathbf{T}^N)\right)$ is the space of all measurable functions $f$ on $\mathbf{R}$ with values in $\mathcal{M}(\mathbf{T}^N)$ such that

$d(f, 0)$ is in $L^\infty(\mathbf{R})$ as function of $t$. The space $L^\infty\left(\mathbf{R}, \mathcal{M}(\mathbf{T}^N)^N\right)$ is defined as $\left(L^\infty\left(\mathbf{R}, \mathcal{M}(\mathbf{T}^N)\right)\right)^N$.

Finally, we expect that the flow map will satisfy the group property, i.e., for any $t, s \in \mathbf{R}$,

$$x(t + s, X) = x\left(t, x(s, X)\right) \quad \text{for a.e.} \quad X. \tag{2.5}$$

---

**Theorem 2.2**
*Assume that (2.1) holds.*

*(i) (Existence) Then there exists a unique $x = x(t, X)$, with*

$$x \in C\left(\mathbf{R}, \left(\mathcal{M}(\mathbf{T}^N)\right)^N\right)$$

*satisfying (2.2)–(2.5). In particular, there exists a renormalized solution to (1.1). Moreover, the mapping $X \mapsto (\beta \circ x)(\cdot, X)$ is in $L^1\left(\mathbf{T}^N, (C(I))^N\right)$ for $\beta$ given in (2.3), (2.4), where $I$ is an arbitrary closed bounded interval. Furthermore, for almost every $X \in \mathbf{T}^N$ the function $t \mapsto x(t, X)$ is in $\left(C^1(\mathbf{R})\right)^N$ and $\dfrac{\partial x}{\partial t} = b(x)$ on $\mathbf{R}$ as a function of $t$.*

*(ii) (Uniqueness) There is at most one (renormalized) solution $x$ to (1.1) satisfying all properties in (i).*

---

It is not difficult to see that the space $C(I)$ is regarded as a Banach space equipped with $\| \cdot \|_\infty$ norm since $I$ is compact.

We shall focus on the uniqueness part of the proof. The main idea to prove the uniqueness is to show that the function $u_0(x(t, X))$ depends only on $u_0 \in C^\infty(\mathbf{T}^N)$ for any choice of a real-valued function $u_0$. Since $u(X, t) = u_0(x(t, X))$ solves a transport equation $u_t - b(X) \cdot \nabla_X u = 0$ with initial datum $u_0(X)$, the problem is reduced to the uniqueness of a (weak) solution to the transport equation with nonsmooth solenoidal coefficient $b$. Here, $u_t = \partial u / \partial t$, and $\nabla_X$ denotes the spatial gradient in $X$. We shall postpone the uniqueness proof of Theorem 2.2 to the end of Chap. 2.

For the reader's convenience, we show that $u(X, t)$ solves

$$u_t(X, t) - b(X) \cdot \nabla_X u(X, t) = 0 \tag{2.6}$$

**Fig. 2.1** Characteristic curve

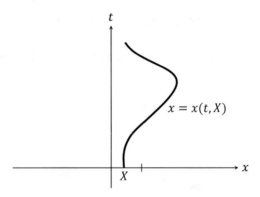

at least if $b \in C^1$ and $x$ is $C^1$ in its variables $(t, X) \in \mathbf{R} \times \mathbf{T}^N$. We first prove that (2.6) at $t = 0$. By direct calculation,

$$u_t(X, t) = \sum_{i=1}^{N} \frac{\partial u_0}{\partial x^i} (x(t, X)) \frac{dx^i}{dt} = \sum_{i=1}^{N} \frac{\partial u_0}{\partial x^i} (x(t, X)) b^i (x(t, X)),$$

$$\frac{\partial u}{\partial X^j}(X, t) = \sum_{i=1}^{N} \frac{\partial u_0}{\partial x^i} (x(t, X)) \frac{\partial x^i}{\partial X^j}(t, X).$$

At $t = 0$, $u_t(X, 0) = \sum_{j=1}^{N} (\partial_j u_0)(X) b^j (X)$, $\partial u / \partial X^j \big|_{t=0} = (\partial_j u_0)(X)$ since $\frac{\partial x^i}{\partial X^j} \big|_{t=0} = \delta_{ij}$ ($\delta_{ij} = 1$ if $i = j$ and $\delta_{ij} = 0$ if $i \neq j$), so we have (2.6). We next set $u^s(X) = u(X, s)$ for $s \in \mathbf{R}$. Then, by the group property, we see that

$$u(X, t + s) = u_0 (x(t + s, X)) = u_0 (x (s, x(t, X))) = u^s (x(t, X)).$$

Applying the result for $t = 0$, with $u_0 = u^s$, we have

$$u_t(X, s) - b(X) \cdot \nabla_X u(X, s) = 0.$$

This yields (2.6). (The curve $x = x(t, X)$ is often called a *characteristic curve* of (2.6). It is easy to see that a solution $u$ of (2.6) is constant along each characteristic curve, i.e., for a fixed $X$, the function $u (x(-t, X), t)$ is constant in $t$; see Fig. 2.1.)

## 2.2    Transport Equations

We are concerned with the uniqueness of a (weak) solution $u = u(x, t)$ to a transport equation

$$u_t - b(x) \cdot \nabla_x u = 0 \tag{2.7}$$

**Fig. 2.2** Support of $\phi$

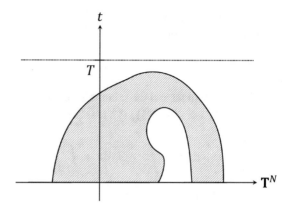

or

$$u_t - \sum_{i=1}^{N} b^i(x) \frac{\partial u}{\partial x^i} = 0,$$

where $b = (b^1, \ldots, b^N)$; here we denote the independent variables by $x$ instead of $X$. A notion of the weak solution $u$ for (2.7) with initial datum $u_0 \in L^1(\mathbf{T}^N)$ is obtained by multiplying $\phi \in C_c^\infty (\mathbf{T}^N \times [0, T))$ (i.e., supp $\phi$ is compact in $\mathbf{T}^N \times [0, T)$) (cf. Fig. 2.2) and integrating over $\mathbf{T}^N \times [0, T]$. Indeed, we have

$$\int_0^T \int_{\mathbf{T}^N} \{\phi u_t - \phi (b(x) \cdot \nabla_x u)\} \, dx dt = 0.$$

Integrating by parts yields

$$-\int_0^T \int_{\mathbf{T}^N} \phi_t u dx dt - \int_{\mathbf{T}^N} \phi u_0 dx + \int_0^T \int_{\mathbf{T}^N} (\text{div}(b\phi)) u \, dx dt = 0. \qquad (2.8)$$

This formula leads to a definition of a weak solution to (2.7). If $\Omega = (0, T)$, then we simply write $L^p(0, T; V)$ instead of $L^p(\Omega, V)$, the space of all $p$th integrable functions on $\Omega$ with values in a Banach space $V$.

---

**Definition 2.3**

Let $b$ be in $W^{1,p'}(\mathbf{T}^N)$, and let $u_0$ be in $L^p(\mathbf{T}^N)$ ($1 \le p \le \infty$). If a function $u \in L^\infty (0, T; L^p(\mathbf{T}^N))$ fulfills (2.8) for all $\phi \in C_c^\infty (\mathbf{T}^N \times [0, T))$, then $u$ is called a *weak solution* to (2.7) with initial datum $u_0$. The vector field $b$ may not be divergence-free. Here $p'$ is the conjugate exponent of $p$, i.e., $1/p + 1/p' = 1$. The integrability conditions guarantee that each term of (2.8) will be well defined as a usual Lebesgue integral. We interpret $1/\infty = 0$ so that $p = 1$ (resp. $p = \infty$) implies $p' = \infty$ (resp. $p' = 1$).

Of course, it is straightforward to extend the definition of a weak solution to an inhomogeneous problem of the form

$$u_t - b \cdot \nabla u = f$$

with initial datum $u_0$ and the inhomogeneous term $f \in L^1\left(0, T; L^1(\mathbf{T}^N)\right)$. We say that $u \in L^\infty\left(0, T; L^p(\mathbf{T}^N)\right)$ is a *weak solution* with initial datum $u_0$ if

$$-\int_0^T \int_{\mathbf{T}^N} \phi_t u \, dx \, dt - \int_{\mathbf{T}^N} \phi_t u_0 \, dx + \int_0^T \int_{\mathbf{T}^N} (\operatorname{div}(b\phi)) \, u \, dx \, dt = \int_0^T \int_{\mathbf{T}^N} f \varphi \, dx \, dt$$

for all $\varphi \in C_c^\infty\left(\mathbf{T}^N \times [0, T)\right)$.

In this section, we simply say that $u$ is a solution to (2.7) if it is a weak solution of (2.8). We are now in a position to state our main uniqueness result.

---

**Theorem 2.4**
*Let $1 \le p \le \infty$, and let $b \in W^{1,p'}(\mathbf{T}^N)$ be solenoidal, i.e., $\operatorname{div} b = 0$ in $\mathbf{T}^N$. Let $u \in L^\infty\left(0, T; L^p(\mathbf{T}^N)\right)$ be a solution to (2.7) with initial datum $u_0 \equiv 0$. Then $u \equiv 0$. (More precisely, $u(x, t) = 0$, a.e. $(x, t) \in \mathbf{T}^N \times (0, T)$.) In particular, if $u_1$ and $u_2$ are solutions to (2.7) with the same initial datum $u_0$, then $u_1 \equiv u_2$ since (2.7) is a linear equation.*

---

A key observation is that $\theta \circ u$ solves (2.7) with initial datum $\theta \circ u_0$ provided that $\theta \in C^1(\mathbf{R})$, with $\theta' \in L^\infty(\mathbf{R})$. This is formally trivial since Eq. (2.7) is linear. However, in our setting, this property is nontrivial. Such a fact is often called a *relabeling lemma*. Here is a rigorous statement in this setting.

---

**Lemma 2.5**
*Let $1 \le p \le \infty$, and let $b \in W^{1,p'}(\mathbf{T}^N)$. Assume that $u \in L^\infty\left(0, T; L^p(\mathbf{T}^N)\right)$ is a solution to (2.7) with initial datum $u_0 \in L^p(\mathbf{T}^N)$. Then $\theta \circ u \in L^\infty\left(0, T; L^p(\mathbf{T}^N)\right)$ is a solution to (2.7) with initial datum $\theta \circ u_0$ provided that $\theta \in C^1(\mathbf{R})$, with $\theta' \in L^\infty(\mathbf{R})$.*

---

Admitting Lemma 2.5, we give a proof of Theorem 2.4 for $1 \le p < \infty$. The case $p = \infty$ is postponed to the next section.

***Proof of Theorem 2.4 for*** $p < \infty$ A heuristic idea is to take $\theta(\sigma) = |\sigma|^p$ and observe that $|u|^p = \theta \circ u$ is a solution to (2.7) so that

$$\frac{d}{dt} \int_{\mathbf{T}^N} |u|^p \, dx = \int_{\mathbf{T}^N} \left( |u|^p \right)_t \, dx = \int_{\mathbf{T}^N} \operatorname{div} \left( b|u|^p \right) \, dx = 0$$

since $\operatorname{div} b = 0$ implies $\operatorname{div}(bw) = b \cdot \nabla w$ for a function $w$. However, since this $\theta$ is not $C^1$ or $\theta' \in L^\infty$, we must circumvent it.

We shall take a suitable $C^1$ function $\theta$. By Lemma 2.5, the function $\theta \circ u$ is a solution of (2.7). Assume that $\theta(0) = 0$. By taking $\phi = \phi(t)$ (spatially constant function) in (2.8), we observe that

$$-\int_0^T \phi_t \left( \int_{\mathbf{T}^N} \theta(u) dx \right) dt - 0 + 0 = 0$$

since $\operatorname{div}(b\phi) = 0$. (This is the only place $\operatorname{div} b = 0$ is invoked.) This implies

$$\int_0^T \psi(t) \left( \int_{\mathbf{T}^N} \theta(u) dx \right) dt = 0$$

for any $\psi \in C_c^\infty((0, T))$ since we are able to take $\phi \in C^\infty([0, T])$ with $\operatorname{supp} \phi \subset [0, T)$ such that $\phi_t = \psi$. Indeed, it suffices to take $\phi = -\int_t^T \psi ds$. Thus a fundamental lemma of the calculus of variations (cf. Exercise 2.3 or [19, Corollary 4.24]) implies that

$$\int_{\mathbf{T}^N} \theta(u)(x, t) dx = 0 \quad \text{for a.e.} \quad t \in (0, T). \tag{2.9}$$

Since $\theta$ is required to be $C^1$ with bounded first derivative, for a given positive constant $M$ we take

$$g(\sigma) := (|\sigma| \wedge M)^p,$$

$$\theta_\varepsilon(\sigma) := (\rho_\varepsilon * g)(\sigma) - (\rho_\varepsilon * g)(0) \quad \text{for} \quad \sigma \in \mathbf{R},$$

where $\rho_\varepsilon \in C_c^\infty(\mathbf{R})$ is a *Friedrichs' mollifier* in $\mathbf{R}$, i.e.,

$$\rho_\varepsilon(\sigma) = \frac{1}{\varepsilon} \rho(\sigma/\varepsilon), \quad \rho \geq 0, \quad \operatorname{supp} \rho \in (-1, 1), \quad \int_{\mathbf{R}} \rho d\sigma = 1$$

for $\varepsilon > 0$; see Fig. 2.3. Here $a_1 \wedge a_2 = \min(a_1, a_2)$ for $a_1, a_2 \in \mathbf{R}$. Since $\theta_\varepsilon \in C^\infty$ (Exercise 2.4), we plug such $\theta_\varepsilon$ into (2.9), and sending $\varepsilon$ to zero (Exercise 2.5) yields

$$\int_{\mathbf{T}^N} (|u| \wedge M)^p (x, t) dx = 0$$

**Fig. 2.3** A typical graph of $\rho$
and $\rho_\varepsilon$

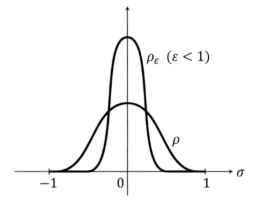

for a.e. $t$ by the (Lebesgue) *dominated convergence theorem* (Theorem 5.1) because $\{|\theta_\varepsilon(u)|\}_{0<\varepsilon<1}$ is bounded in $L^\infty(\mathbf{T}^N)$. Since $u$ is in $L^\infty\left(0, T; L^p(\mathbf{T}^N)\right)$, we send $M$ to infinity and again use the dominated convergence theorem to conclude that

$$\int_{\mathbf{T}^N} |u|^p(x, t)\mathrm{d}x = 0, \quad \text{a.e.} \quad t \in (0, T).$$

Thus, we conclude that $u \equiv 0$ in $L^\infty\left(0, T; L^p(\mathbf{T}^N)\right)$. By Fubini's theorem, this is simply $u(x, t) = 0$ for a.e. $(x, t) \in \mathbf{T}^N \times (0, T)$.    □

In the rest of this section, we shall prove the relabeling lemma (Lemma 2.5). A key idea is an approximation. From here, let $\rho_\varepsilon$ be a Friedrichs' mollifier in $\mathbf{R}^N$, i.e., for $\varepsilon > 0$

$$\rho_\varepsilon(x) = \frac{1}{\varepsilon^N}\rho\left(\frac{x}{\varepsilon}\right), \quad \rho \in C_c^\infty(\mathbf{R}^N), \quad \int_{\mathbf{R}^N} \rho\,\mathrm{d}x = 1, \quad \rho \geq 0.$$

**Lemma 2.6**

*Let $1 \leq p \leq \infty$, and let $b \in W^{1,\beta}(\mathbf{T}^N)$, with $\beta \geq p'$. Let $u \in L^\infty\left(0, T; L^p(\mathbf{T}^N)\right)$ be a solution of (2.7) with initial datum $u_0 \in L^p(\mathbf{T}^N)$. Let $\rho_\varepsilon$ be a Friedrichs' mollifier in space variables. Then $u_\varepsilon = u * \rho_\varepsilon$ satisfies*

$$\frac{\partial u_\varepsilon}{\partial t} - b \cdot \nabla u_\varepsilon = r_\varepsilon$$

*with initial datum $u_{0\varepsilon} = u_0 * \rho_\varepsilon$, with some real-valued function $r_\varepsilon$ converging to zero in $L^1\left(0, T; L^\alpha(\mathbf{T}^N)\right)$ as $\varepsilon \to 0$, where*

$$1/\alpha = 1/\beta + 1/p \text{ if } \beta \text{ or } p \text{ is finite,}$$

$$1 \leq \alpha < \infty \text{ is arbitrary if } \beta = p = \infty.$$

***Proof of Lemma 2.5 admitting Lemma 2.6***  By Lemma 2.6, we observe that

$$\frac{\partial u_\varepsilon}{\partial t} - b \cdot \nabla u_\varepsilon = r_\varepsilon \to 0 \quad \text{in} \quad L^1\left(0, T; L^1(\mathbf{T}^N)\right);$$

in other words,

$$\lim_{\varepsilon \downarrow 0} \int_0^T \int_{\mathbf{R}^N} |r_\varepsilon| \, dx dt = 0.$$

We take $\theta \in C^1(\mathbf{R})$, with $\theta' \in L^\infty(\mathbf{R})$. Since $u_\varepsilon$ is smooth in space, we see that

$$\frac{\partial}{\partial t}(\theta \circ u_\varepsilon) - b \cdot \nabla(\theta \circ u_\varepsilon) = r_\varepsilon \theta' \circ u_\varepsilon.$$

Since $\theta' \in L^\infty(\mathbf{R})$, the right-hand side $r_\varepsilon \theta' \circ u_\varepsilon \to 0$ in $L^1\left(0, T; L^1(\mathbf{R}^N)\right)$ as $\varepsilon \downarrow 0$, we formally conclude that $\theta \circ u$ is a solution to (2.7) with initial datum $\theta \circ u_0$. Of course, we must carry out these arguments in a weak form, (2.8). In other words, we send $\varepsilon \downarrow 0$ for

$$-\int_0^T \int_{\mathbf{T}^N} \phi_t (\theta \circ u_\varepsilon) \, dx dt - \int_{\mathbf{T}^N} \phi (\theta \circ u_{0\varepsilon}) \, dx + \int_0^T \int_{\mathbf{T}^N} (\mathrm{div}(b\phi)) \, \theta \circ u^\varepsilon \, dx dt$$

$$= \int_0^T \int_{\mathbf{T}^N} \phi r_\varepsilon (\theta' \circ u_\varepsilon) \, dx dt \quad \text{with} \quad u_{0\varepsilon} = u_0 * \rho_\varepsilon$$

to get

$$-\int_0^T \int_{\mathbf{T}^N} \phi_t (\theta \circ u_\varepsilon) \, dx dt - \int_{\mathbf{T}^N} \phi (\theta \circ u_0) \, dx + \int_0^T \int_{\mathbf{T}^N} (\mathrm{div}(b\phi)) \, \theta \circ u^\varepsilon \, dx dt = 0$$

for $\phi \in C_c^\infty \left(\mathbf{T}^N \times [0, T)\right)$.  $\square$

It remains to prove Lemma 2.6. For this purpose, it suffices to prove the convergence of commutators.

---

**Lemma 2.7**

*Let $\rho_\varepsilon$ denote a Friedrichs' mollifier in $\mathbf{R}^N$. Let $1 \leq p \leq \infty$, and let $b \in W^{1,\beta}(\mathbf{T}^N)$, with $\beta \geq p'$. If $w \in L^p(\mathbf{T}^N)$, then*

$$R_\varepsilon(w, b) = (b \cdot \nabla w) * \rho_\varepsilon - b \cdot \nabla(w * \rho_\varepsilon) \to 0 \quad \text{in} \quad L^\alpha(\mathbf{T}^N)$$

*as $\varepsilon \to 0$, where $\alpha$ is given as in Lemma 2.6; in particular, in the case $p = \infty$ so that $p' = 1$, $\alpha = \beta$ if $\beta < \infty$. Moreover, $\|R_\varepsilon(w, b)\|_{L^\alpha} \leq C\|w\|_{L^p}\|Db\|_{L^\beta}$ with some $C > 0$ independent of sufficiently small $\varepsilon$.*

**Proof of Lemma 2.6 admitting Lemma 2.7** Direct calculation shows that

$$\frac{\partial}{\partial t} u_\varepsilon - b \cdot \nabla u_\varepsilon = R_\varepsilon(u, b);$$

in other words,

$$-\int_0^T \int_{\mathbf{T}^N} \phi_t u_\varepsilon \, dx dt - \int_{\mathbf{T}^T} \phi u_{0\varepsilon} \, dx + \int_0^T \int_{\mathbf{T}^N} (\text{div}(b\phi)) u_\varepsilon \, dx dt$$

$$= \int_0^T \int_{\mathbf{T}^N} R_\varepsilon(u, b) \phi \, dx dt$$

for all $\phi \in C_c^\infty \left( \mathbf{T}^N \times [0, T) \right)$. By the second statement of Lemma 2.7, we have

$$\| R_\varepsilon(u, b) \|_{L^\alpha}(t) \le C \|u\|_{L^p}(t) \|Db\|_{L^\beta}.$$

The right-hand side is bounded in $t$, so Lemma 2.7 yields

$$\lim_{\varepsilon \downarrow 0} \int_0^T \| R_\varepsilon(w, b) \|_{L^\alpha}(t) \, dt = \int_0^T \lim_{\varepsilon \downarrow 0} \| R_\varepsilon(w, b) \|_{L^\alpha}(t) \, dt = 0$$

by a dominated convergence theorem. (One immediately observes that $b$ is allowed to depend on time $t$ if $b \in L^1 \left( 0, T; W^{1,\beta}(\mathbf{T}^N) \right)$ and $u \in L^\infty \left( 0, T; L^p(\mathbf{T}^N) \right)$. □

**Proof of Lemma 2.7** We first observe that the term $(b \cdot \nabla w) * \rho_\varepsilon$ should be interpreted as

$$((b \cdot \nabla w) * \rho_\varepsilon)(x) = -\int_{\mathbf{T}^N} w(y) \, \text{div}_y \{b(y) \rho_\varepsilon(x - y)\} \, dy$$

since $\nabla w$ is not an integrable function. This identity is easily obtained if $w$ is smooth by integration by parts. We proceed with

$$(b \cdot \nabla w) * \rho_\varepsilon - b \cdot \nabla (w * \rho_\varepsilon) = \rho_\varepsilon * (b \cdot \nabla w) - \sum_{i=1}^N \left( \frac{\partial}{\partial x_i} \rho_\varepsilon * w \right) b^i$$

$$= -\int_{\mathbf{T}^N} w(y) \left[ \text{div}_y \{b(y)\rho_\varepsilon(x - y)\} + b(x) \cdot (\nabla \rho_\varepsilon)(x - y) \right] dy$$

$$= \int_{\mathbf{T}^N} w(y) ((b(y) - b(x)) \cdot (\nabla \rho_\varepsilon)(x - y)) \, dy - \int_{\mathbf{T}^N} (w \, \text{div} \, b) * \rho_\varepsilon \, dy$$

$$= I + II.$$

Here, we use the relation $\nabla_y (\rho_\varepsilon(x - y)) = -(\nabla \rho_\varepsilon)(x - y)$.

We next estimate $I$. Since $\rho$ has compact support, $\rho_\varepsilon(x) = 0$ for $|x| \geq C\varepsilon$, with some $C$ independent of $\varepsilon$. By changing $\varepsilon$ by $c_0\varepsilon$ with some $c_0 > 0$, we may assume that $C < \min_i \omega_i/2$. This choice implies that the ball $B_C(z)$ centered at $z \in \mathbf{T}^N$ with radius $C$ is contained in one periodic cell. We shall assume that $\varepsilon < 1$. By changing the variable of integration by $z = (y - x)/\varepsilon$, we obtain

$$I(x) = \int_{|x-y| \leq C\varepsilon} \{(b(y) - b(x)) \cdot (\nabla\rho_\varepsilon)(x - y)w(y)\}\, dy$$

$$= \int_{|z| \leq C} \left\{ \frac{b(x + \varepsilon z) - b(x)}{\varepsilon} \cdot (\nabla\rho)(-z)w(x + \varepsilon z) \right\} dz.$$

Since $\nabla\rho$ is bounded, i.e., $C_0 := \|\nabla\rho\|_{L^\infty(\mathbf{R}^N)} < \infty$, we observe that

$$|I(x)| \leq C_0 \int_{|z| \leq C} k_\varepsilon(x, z)\, |w(x + \varepsilon z)|\, dz$$

with $k_\varepsilon(x, z) = |b(x + \varepsilon z) - b(x)|/\varepsilon$. By the *Hölder inequality*

$$\|f \cdot 1\|_{L^1(B_C)}^\alpha \leq \|f\|_{L^\alpha(B_C)}^\alpha |B_C|^{\alpha-1},$$

we observe that

$$\frac{1}{C_0^\alpha}\|I\|_{L^\alpha(\mathbf{T}^N)}^\alpha = \int_{\mathbf{T}^N} \left\{ \int_{|z| \leq C} k_\varepsilon(x, z)\, |w(x + \varepsilon z)|\, dz \right\}^\alpha dx$$

$$\leq \int_{\mathbf{T}^N} \int_{|z| \leq C} \{k_\varepsilon(x, z)\, |w(x + \varepsilon z)|\}^\alpha\, dz dx |B_C|^{\alpha-1} =: J^\alpha$$

for $\alpha \in [1, \infty]$. Here, $|B_C| = \mathcal{L}^N(B_C)$ (i.e., the volume of a ball of radius $C$) and it equals $|S^{N-1}|C^N/N$. Applying the Hölder inequality for $k_\varepsilon^\alpha$ and $|w|^\alpha$ with $1/\alpha = 1/\beta + 1/p$, we see that

$$J \leq |B_C|^{1-1/\alpha} \left\{ \int_{\mathbf{T}^N} \int_{|z| \leq C} k_\varepsilon(x, z)^\beta dz dx \right\}^{1/\beta} \left\{ \int_{\mathbf{T}^N} \int_{|z| \leq C} |w(x + \varepsilon z)|^p dz dx \right\}^{1/p}$$

$$= C_1\|w\|_{L^p(\mathbf{T}^N)} \left\{ \int_{\mathbf{T}^N} \int_{|z| \leq C} k_\varepsilon(x, z)^\beta dz dx \right\}^{1/\beta}$$

with $C_1 = |B_C|^{1-1/\alpha}|B_C|^{1/p} = |B_C|^{1-1/\beta}$. Since

$$|b(x + \varepsilon z) - b(x)| = \left\{ \sum_{i=1}^{N} \left| \int_0^1 \left\langle \nabla b^i(x + \varepsilon \sigma z), \varepsilon z \right\rangle d\sigma \right|^2 \right\}^{1/2}$$

$$\leq \varepsilon |z| \left( \sum_{i=1}^{N} \left| \int_0^1 \left| \nabla b^i(x + \varepsilon \sigma z) \right| d\sigma \right|^2 \right)^{1/2}$$

$$\leq \varepsilon |z| \int_0^1 |Db(x + \varepsilon \sigma z)| \, d\sigma,$$

we see that

$$\int_{\mathbf{T}^N} \int_{|z| \leq C} k_\varepsilon(x, z)^\beta dz dx \leq \int_{\mathbf{T}^N} \int_{|z| \leq C} \int_0^1 |Db(x + \varepsilon \sigma z)|^\beta \, d\sigma |z|^\beta dz dx$$

$$= C_2^\beta \|Db\|_{L^\beta(\mathbf{T}^N)}^\beta \quad \text{with} \quad C_2 = \left( \int_{|z| \leq C} |z|^\beta dz \right)^{1/\beta}.$$

Here, $|Db(x)|$ denotes the Euclidean norm of the $N \times N$ matrix $Db(x)$ in $\mathbf{R}^{N \times N}$. In other words, $|Db(x)|^2 = \mathrm{tr}\left(Db(x)Db(x)^T\right)$, where $M^T$ denotes the transpose of a matrix $M$.

We now conclude that

$$\|I\|_{L^\alpha(\mathbf{T}^N)} \leq C_0 J \leq C_0 C_1 \|w\|_{L^p(\mathbf{T}^N)} \left\{ \int_{\mathbf{T}^N} \int_{|z| \leq C} k_\varepsilon(x, z)^\beta dz dx \right\}^{1/\beta}$$

$$\leq C_0 C_1 C_2 \|w\|_{L^p(\mathbf{T}^N)} \|Db\|_{L^\beta(\mathbf{T}^N)}.$$

By *Young's inequality* for convolution, we have

$$\|II\|_{L^\alpha} \leq \|\rho_\varepsilon\|_{L^1} \|w \, \mathrm{div} \, b\|_{L^\alpha} \leq 1 \|w \, \mathrm{div} \, b\|_{L^\alpha}.$$

By the Hölder inequality,

$$\|w \, \mathrm{div} \, b\|_{L^\alpha} \leq \|w\|_{L^p} \|Db\|_{L^\beta}.$$

Thus, the desired estimate $\|R_\varepsilon(w, b)\|_{L^\alpha} \leq C \|w\|_{L^p} \|Db\|_{L^\beta}$ now follows.

It remains to prove that $\|R_\varepsilon(w, b)\|_{L^\alpha} \to 0$ as $\varepsilon \to 0$. This can be carried out by a density argument.

If $w \in W^{1,p}(\mathbf{T}^N)$, then both $(b \cdot \nabla w) * \rho_\varepsilon$ and $(b \cdot \nabla)(w * \rho_\varepsilon)$ converge to $(b \cdot \nabla)w$ in $L^\alpha(\mathbf{T}^N)$ as $\varepsilon \to 0$ since $f * \rho_\varepsilon \to f$ in $L^\alpha(\mathbf{T}^N)$ if $f \in L^\alpha(\mathbf{T}^N)$; see [19, Section 4.4]. Here, we invoke the property $\alpha < \infty$. Thus, $R_\varepsilon(w, b) \to 0$ in $L^\alpha(\mathbf{T}^N)$ provided that $w \in W^{1,p}(\mathbf{T}^N)$.

Suppose that $1 \leq p < \infty$. Since $W^{1,p}(\mathbf{T}^N)$ is *dense* in $L^p(\mathbf{T}^N)$ for $p < \infty$, for $w \in L^p(\mathbf{T}^N)$ there is a sequence $\{w_m\} \subset W^{1,p}(\mathbf{T}^N)$ converging to $w$ in $L^p(\mathbf{T}^N)$. (This density follows from the fact that $f * \rho_{1/m} \in C^\infty(\mathbf{T}^N)$ and $f * \rho_{1/m} \to f$ in $L^p(\mathbf{T}^N)$ as $m \to \infty$ for $p \in [1, \infty)$. See [19, Section 4.4] and Appendix 5.2 (6).) Since

$$R_\varepsilon(w, b) = R_\varepsilon(w - w_m, b) + R_\varepsilon(w_m, b),$$

we observe that

$$\|R_\varepsilon(w, b)\|_{L^\alpha} \leq C_3 \|w - w_m\|_{L^p} \|Db\|_{L^\beta} + \|R_\varepsilon(w_m, b)\|_{L^\alpha}$$

by the estimate of Lemma 2.7 proved earlier with some $C_3 > 0$ independent of sufficiently small $\varepsilon > 0$. Sending $\varepsilon \downarrow 0$ yields

$$\limsup_{\varepsilon \downarrow 0} \|R_\varepsilon(w, b)\|_{L^\alpha} \leq C_3 \|w - w_m\|_{L^p} \|Db\|_{L^\beta} + 0.$$

Letting $m \to \infty$ yields a desired conclusion, i.e., $R_\varepsilon(w, b) \to 0$ as $\varepsilon \to 0$ in $L^\alpha(\mathbf{T}^N)$.

If $p = \beta = \infty$, then, by our assumption, $\alpha < \infty$. This case is reduced to the case $\alpha = p < \infty$, $\beta = \infty$.

It remains to prove the case $\beta < \infty$, but $p = \infty$. In this case, $\alpha = \beta$. Unfortunately, $w \in L^\infty(\mathbf{T}^N)$ cannot be approximated by an element of $W^{1,\infty}(\mathbf{T}^N)$ in the $L^\infty$ sense. However, it is still possible to approximate in a weaker sense. That is, for any $w \in L^\infty(\mathbf{T}^N)$, there exists a sequence $w_n \in L^\infty(\mathbf{T}^N)$ such that $w_m \to w$ a.e. and $\|w_m\|_\infty \leq \|w\|_\infty$. (For example, it is enough to take $w_m = \rho_{1/m} * w$. See Exercise 2.10.) We are able to estimate

$$\limsup_{\varepsilon \to 0} \|R_\varepsilon(v, b)\|_{L^\alpha}^\alpha \leq C \int_{\mathbf{T}^N} |Db(x)|^\alpha \, |v(x)|^\alpha \, dx \qquad (2.10)$$

with some $C$ independent of $v \in L^\infty(\mathbf{T}^N)$ and $b$. Indeed, from a similar argument as previously, the term corresponding to $I$ is dominated by

$$\frac{1}{C_0^\alpha} \|I\|_{L^\alpha(\mathbf{T}^N)}^\alpha \leq \int_{\mathbf{T}^N} \int_{|z| \leq C} \{k_\varepsilon(x, z) \, |v(x + \varepsilon z)|\}^\alpha \, dz dx |B_C|^{\alpha - 1}$$

$$\leq \int_{\mathbf{T}^N} \int_{|z| \leq C} |z|^\alpha \left( \int_0^1 |Db(x + \varepsilon \sigma z)|^\alpha \, |v(x + \varepsilon z)|^\alpha \, d\sigma \right) dz dx |B_C|^{\alpha - 1},$$

where $\beta = \alpha$. We change the variable of integration by $\overline{x} = x + \varepsilon \sigma z$ to get

$$= \int_{\mathbf{T}^N} \int_{|z| \leq C} |z|^\alpha \int_0^1 |Db(\overline{x})|^\alpha \, |v\,(\overline{x} + (1 - \sigma)\varepsilon z)|^\alpha \, d\sigma dz d\overline{x} |B_C|^{\alpha - 1}.$$

Since the shift operator is continuous in $L^\alpha$ norm,[1] we see that

$$v\left(\overline{x} + (1-\sigma)\varepsilon z\right) \to v(\overline{x})$$

for almost all $\overline{x} \in \mathbf{T}^N$ as $\varepsilon \to 0$ by taking a subsequence if necessary.[2] Since $v$ is bounded and $|Db|^\alpha$ is integrable, by the dominated convergence theorem we conclude that the last term converges to

$$C_2^\alpha |B_C|^{\alpha-1} \int_{\mathbf{T}^N} |Db(x)|^\alpha\, |v(x)|^\alpha\, dx \quad \text{as} \quad \varepsilon \downarrow 0.$$

A similar but much easier observation yields a similar estimate for $II$. We thus conclude (2.10).

By (2.10), we are able to estimate

$$\limsup_{\varepsilon \downarrow 0} \|R_\varepsilon(w,b)\|_{L^\alpha} \le \limsup_{\varepsilon \downarrow 0} \|R_\varepsilon(w-w_m,b)\|_{L^\alpha}$$

$$\le C \int_{\mathbf{T}^N} |Db(x)|^\alpha\, |w(x) - w_m(x)|^\alpha\, dx.$$

Since $\|w_m\|_\infty \le \|w\|_\infty$ and $w_m \to w$ a.e. as $m \to \infty$, by the dominated convergence theorem, we conclude that the right-hand side tends to zero. The proof for the convergence $\lim_{\varepsilon \downarrow 0} \|R_\varepsilon(w,b)\|_{L^\alpha} = 0$ is now complete. $\qquad\square$

## 2.3    Duality Argument

In this section, we shall prove the uniqueness result (Theorem 2.4) of a solution to the transport equation (2.7) when it is bounded under the condition that the coefficient $b$ of the transport term is merely in $W^{1,1}(\mathbf{T}^N)$. The argument presented so far does not work for $p = \infty$. We study the case where $p = \infty$ by a duality argument.

Let us explain a basic idea of the duality argument. This is a typical argument for uniqueness. Consider a linear operator $S$ from $\mathbf{R}^n$ to $\mathbf{R}^m$. Suppose that we are asked to check whether this mapping is injective or one to one. Since $S$ is linear, it is enough to show the kernel of $S$ is just $\{0\}$. In other words, we are asked to prove that $Sx = 0$ implies $x = 0$. The main idea of the duality argument is to reduce the

---

[1] This is one of the fundamental properties of the Lebesque measure. It states that $\lim_{|y|\to 0} \|\tau_y f - f\|_{L^\alpha(\mathbf{T}^N)} = 0$ for $\alpha \in [1,\infty)$, where $(\tau_y f)(x) = f(x+y)$.

[2] If $f_\varepsilon \to f$ in $L^p(\mathbf{T}^N)$, there is a subsequence $f_{\varepsilon_k}$ that converges to $f$ a.e.

problem to the solvability of its dual problem $S^*z = y$ for all $y \in \mathbf{R}^n$. If there is a solution $z$, then

$$x \cdot y = x \cdot S^*z = Sx \cdot z = 0$$

for all $y \in \mathbf{R}^n$. This implies $x = 0$.

So to carry out this argument, we need some existence theorem for a dual problem.

---

**Proposition 2.8**

*Let $1 \leq p \leq \infty$, and let $u_0 \in L^p(\mathbf{T}^N)$. Assume that $b \in W^{1,1}(\mathbf{T}^N) \cap L^{p'}(\mathbf{T}^N)$, with* $\operatorname{div} b = 0$. *Then there exists a solution $u \in L^\infty\left(0, T; L^p(\mathbf{T}^N)\right)$ of (2.7) with initial datum $u_0$. Here, $1/p + 1/p' = 1$.*

---

*Proof.* A typical way to prove the existence of a solution under nonsmooth coefficients is as follows. We first approximate the problem by regularization, then take a limit of a solution to the approximate problem. We need a priori estimates to carry out the second step.

We begin with an a priori estimate assuming that $b$ and $u_0$ are smooth. Let $x = x(t, X)$ be the flow map generated by $b$, i.e., $\dot{x} = b(x)$, with $x(0) = X$. If $b$ is smooth, then by the smooth dependence of the initial data [52, Chapter 5], $x$ is smooth in $t$ and $X$. Moreover, by the uniqueness of the solution of (1.1) (Proposition 1.1), the group property (2.5) holds. We first recall that $u(X, t) = u_0\left(x(t, X)\right)$ (uniquely) solves (2.7), i.e.,

$$u_t - b(X) \cdot \nabla_X u = 0$$

with smooth data

$$u|_{t=0}(X) = u_0(X)$$

if $b$ is smooth. By this solution formula, it is clear that

$$\|u\|_{L^\infty}(t) \leq \|u_0\|_{L^\infty}.$$

By the group property (2.5), we have $u_0(X) = u\left(x(-t, X), t\right)$. Thus, $\|u_0\|_{L^\infty} \leq \|u\|_{L^\infty}(t)$, so that

$$\|u\|_{L^\infty}(t) = \|u_0\|_{L^\infty}.$$

Here, $\|u\|_{L^q}(t)$ denotes the norm of $u$ in $L^q(\mathbf{T}^N)$ with a parameter $t$. By a formal argument, to obtain the volume-preserving property (2.2), we observed that the

Jacobian det $F$ of the flow map equals one, where $F = (F_{ij}) = (\partial x^i / \partial X^j)$. Thus, by (2.2), we see that

$$\|u\|_{L^q}(t) = \|u_0\|_{L^q} \tag{2.11}$$

for $1 \le q < \infty$. Combining the $L^\infty$ estimate, we see (2.11) holds for all $q \in [1, \infty]$.

We next approximate our original $b$ by $b_\varepsilon = b * \rho_\varepsilon$. Then there is a (unique) solution $u_\varepsilon(x, t) \in C^\infty \left( \mathbf{T}^N \times [0, T) \right)$ of

$$u_t - b_\varepsilon(x) \cdot \nabla_x u = 0 \tag{2.12}$$

with initial datum $u_{0\varepsilon} = u_0 * \rho_\varepsilon$ for $u_0 \in L^P(\mathbf{T}^N)$; the solution $u_\varepsilon(x, t)$ is given by $u_\varepsilon(x, t) = u_0 (x_\varepsilon(t, x))$. Here, $x_\varepsilon$ is the flow map generated by $b_\varepsilon$, i.e., $\dot{x}_\varepsilon = b_\varepsilon(x_\varepsilon)$, with $x_\varepsilon(0) = x \in \mathbf{T}^N$. Since $u_\varepsilon$ solves (2.12), it solves its weak form, i.e.,

$$-\int_0^T \int_{\mathbf{T}^N} \phi_t u_\varepsilon \mathrm{d}x \mathrm{d}t - \int_{\mathbf{T}^N} \phi u_{0\varepsilon} \mathrm{d}x + \int_0^T \int_{\mathbf{T}^N} (\mathrm{div}(b_\varepsilon \phi)) u_\varepsilon \mathrm{d}x \mathrm{d}t = 0 \tag{2.13}$$

for all $\phi \in C_c^\infty \left( \mathbf{T}^N \times [0, T) \right)$. We note that $u_\varepsilon$ and $u_{0\varepsilon}$ satisfy the norm-preserving property (2.11).

Case 1 ($1 < p \le \infty$). Since (2.11) for $u_\varepsilon$ implies that $\{u_\varepsilon\}_{0<\varepsilon<1}$ is bounded in $L^\infty \left( 0, T; L^P(\mathbf{T}^N) \right)$, by $*$-weak compactness, there is a subsequence $\{u_{\varepsilon'}\}$ converging to some $u$ $*$-weakly in $L^\infty \left( 0, T; L^P(\mathbf{T}^N) \right)$ for $p \in (1, \infty]$; see Appendix 5.2 (4) for $*$-weak convergence in $L^\infty \left( 0, T; L^P(\mathbf{T}^N) \right)$. We now send $\varepsilon'$ to zero in (2.13). It is easy to see that the first two terms of (2.13) converge to the first two terms of (2.8), respectively. The only difficulty lies in handling the last term. We proceed with

$$\int_0^T \int_{\mathbf{T}^N} \mathrm{div}(b_{\varepsilon'} \phi) u_{\varepsilon'} \mathrm{d}x \mathrm{d}t = \int_0^T \int_{\mathbf{T}^N} (b_{\varepsilon'} \cdot \nabla \phi) u_{\varepsilon'} \mathrm{d}x \mathrm{d}t \tag{2.14}$$

since $\mathrm{div}\, b_\varepsilon = \mathrm{div}(b * \rho_\varepsilon) = (\mathrm{div}\, b) * \rho_\varepsilon = 0$. By a standard property of the mollifier, we see that $b_\varepsilon \to b$ in $L^{p'}(\mathbf{T}^N)$ if $p \in (1, \infty]$ (e.g., [19, Section 4.4]). Since $u_{\varepsilon'} \rightharpoonup u$ $*$-weakly in $L^\infty \left( 0, T; L^P(\mathbf{T}^N) \right)$, this implies that (2.14) converges to

$$\int_0^T \int_{\mathbf{T}^N} (b \cdot \nabla \phi) u \, \mathrm{d}x \mathrm{d}t = \int_0^T \int_{\mathbf{T}^N} \mathrm{div}(b \phi) u \, \mathrm{d}x \mathrm{d}t$$

as $\varepsilon' \downarrow 0$. Here, we invoked the property

$$\lim_{\varepsilon \downarrow 0} \int_0^T \int_{\mathbf{T}^N} f_\varepsilon g_\varepsilon \, \mathrm{d}x \mathrm{d}t = \int_0^T \int_{\mathbf{T}^N} f g \, \mathrm{d}x \mathrm{d}t$$

if $f_\varepsilon \to f$ in $L^1\left(0, T; L^p(\mathbf{T}^N)\right)$ and $g_\varepsilon \rightharpoonup g$ *-weakly in $L^\infty\left(0, T; L^p(\mathbf{T}^N)\right)$ as $\varepsilon \downarrow 0$. To see this property, we notice

$$f_\varepsilon g_\varepsilon - fg = (f_\varepsilon - f)g_\varepsilon + f(g_\varepsilon - g)$$

so that

$$\left| \int_0^T \int_{\mathbf{T}^N} f_\varepsilon g_\varepsilon \, dx dt - \int_0^T \int_{\mathbf{T}^N} fg \, dx dt \right|$$

$$\leq \|f_\varepsilon - f\|_{L^1\left(0,T;L^{p'}(\mathbf{T}^N)\right)} \|g_\varepsilon\|_{L^\infty(0,T;L^p(\mathbf{T}^N))}$$

$$+ \left| \int_0^T \int_{\mathbf{T}^N} f(g_\varepsilon - g) \, dx dt \right|$$

by the Hölder inequality (cf. Exercise 2.9). The first term tends to zero as $\varepsilon \downarrow 0$ since $\|g_\varepsilon\|_{L^\infty(0,T;L^p(\mathbf{T}^N))}$ is bounded and $f_\varepsilon \to f$ in $L^1\left(0, T; L^{p'}(\mathbf{T}^N)\right)$. The second term tends to zero as $\varepsilon \downarrow 0$ since $g_\varepsilon \rightharpoonup g$ *-weakly in $L^\infty\left(0, T; L^p(\mathbf{T}^N)\right)$. We thus obtain (2.8) when $p \in (1, \infty]$.

Case 2 ($p = 1$). In this case, boundedness in $L^1(\mathbf{T}^N)$ does not imply weak compactness. We approximate $u_0 \in L^1(\mathbf{T}^N)$ by $u_{0m} \in L^{\hat{p}}(\mathbf{T}^N)$, $\hat{p} > 1$ such that $\|u_0 - u_{0m}\|_{L^1} \to 0$ as $m \to \infty$. Let $u_m^\varepsilon$ be an $L^\infty\left(0, T; L^{\hat{p}}(\mathbf{T}^N)\right)$ solution with initial datum $u_{0m}$ for (2.12), which is given by

$$u_m^\varepsilon(x, t) = u_{0m}\left(x_\varepsilon(t, x)\right).$$

Since (2.12) is linear, we apply (2.11) to $u_m^\varepsilon - u_{m+1}^\varepsilon$ to get

$$\|u_m^\varepsilon - u_{m+1}^\varepsilon\|_{L^q}(t) = \|u_{0m} - u_{0m+1}\|_{L^q} \tag{2.15}$$

for all $1 \leq q \leq \hat{p}$. For $m = 1$, we take a subsequence as $\varepsilon \to 0$ to get $U_1$ satisfying (2.8) starting with $u_{01}$ by Case 1. For $m = 2$, we take a further subsequence to get a $U_2$ satisfying (2.8) starting with $u_{02}$. We repeat the procedure and obtain $U_m \in L^\infty\left(0, T; L^{\hat{p}}(\mathbf{T}^N)\right)$ satisfying (2.8) with initial datum $u_{0m}$ and moreover satisfying

$$\|U_m - U_{m+1}\|_{L^1}(t) = \|u_{0m} - u_{0m+1}\|_{L^1}$$

by (2.15). This implies that $\{U_m\}$ is a *Cauchy sequence* in $L^\infty\left(0, T; L^1(\mathbf{T}^N)\right)$. Since $U_m$ solves (2.8), letting $m \to \infty$ yields the desired solution $U = \lim_{m\to\infty} U_m \in L^\infty\left(0, T; L^1(\mathbf{T}^N)\right)$ of (2.7) with initial datum $u_0 \in L^1(\mathbf{T}^N)$. $\quad\square$

▶ **Remark 2.9**

**(i)** The assumption $b \in W^{1,1}(\mathbf{T}^N)$ is used only to define div $b$ in $L^1(\mathbf{T}^N)$. It is enough to assume that $b \in L^{p'}(\mathbf{T}^N)$ with div $b = 0$ in the distribution sense.

**(ii)** Proposition 2.8 is still valid for a solution of $u_t - b \cdot \nabla u = f$ (instead of (2.7)) provided that $f \in L^1\left(0, T; L^p(\mathbf{T}^N)\right)$. Its weak form is given right after Definition 2.3.

***Proof of Theorem 2.4 for*** $p = \infty$   We shall prove that

$$\int_0^T \int_{\mathbf{T}^N} u\phi \mathrm{d}x \mathrm{d}t = 0$$

for all $\phi \in C_c^\infty\left(\mathbf{T}^N \times (0, T)\right)$. This implies $u = 0$ a.e. in $\mathbf{T}^N \times (0, T)$ by a fundamental lemma of the calculus of variations (Exercise 2.3 and [19, Corollary 4.24]).

We first consider a dual problem, which is a backward problem:

$$\frac{\partial \Phi}{\partial t} - b \cdot \nabla\Phi = \phi \text{ in } \mathbf{T}^N \times (0, T), \quad \Phi|_{t=T} = 0 \text{ in } \mathbf{T}^N.$$

By Proposition 2.8 and Remark 2.9 (ii), there exists a solution

$$\tilde{\Phi} \in L^\infty\left(0, T; L^\infty(\mathbf{T}^N)\right)$$

for $\partial_t \tilde{\Phi} + b \cdot \nabla \tilde{\Phi} = \tilde{\phi}$ with $\tilde{\Phi}(x, 0) = 0$ for $\tilde{\phi}(x, t) = \phi(x, T - t)$. Setting $\Phi(x, t) = \tilde{\Phi}(x, T-t)$, we find a solution $\Phi \in L^\infty\left(0, T; L^\infty(\mathbf{T}^N)\right)$ to the preceding backward problem.

We regularized $\Phi$ and $u$ by $\Phi_\varepsilon = \Phi * \rho_\varepsilon$ and $u_\varepsilon = u * \rho_\varepsilon$, respectively. The resulting equation for $\Phi_\varepsilon$ and $u_\varepsilon$ is

$$\frac{\partial \Phi_\varepsilon}{\partial t} - b \cdot \nabla\Phi_\varepsilon = \phi_\varepsilon + \psi_\varepsilon \text{ in } \mathbf{T}^N \times (0, T), \qquad \Phi_\varepsilon|_{t=T} = 0 \text{ in } \mathbf{T}^N,$$

$$\frac{\partial u_\varepsilon}{\partial t} - b \cdot \nabla u_\varepsilon = r_\varepsilon \text{ in } \mathbf{T}^N \times (0, T), \qquad\qquad u_\varepsilon|_{t=0} = 0 \text{ in } \mathbf{T}^N$$

with $\phi_\varepsilon = \phi * \rho_\varepsilon$, $\psi_\varepsilon = (b \cdot \nabla\Phi) * \rho_\varepsilon - b \cdot ((\nabla\Phi) * \rho_\varepsilon)$ and

$$r_\varepsilon = (b \cdot \nabla u) * \rho_\varepsilon - b \cdot ((\nabla u) * \rho_\varepsilon).$$

We have $r_\varepsilon, \psi_\varepsilon \to 0$ (as $\varepsilon \downarrow 0$) in $L^1\left(0, T; L^1(\mathbf{T}^N)\right)$ by Lemma 2.6; the external term $\phi$ is also allowed in Lemma 2.6. Multiply the second equation by $\Phi_\varepsilon$; integrating parts yields

$$-\int_0^T \int_{\mathbf{T}^N} \{u_\varepsilon (\phi_\varepsilon + \psi_\varepsilon) + r_\varepsilon \Phi_\varepsilon\}\, dxdt = 0;$$

here we invoke the property that div $b = 0$. The term $\int_0^T \int_{\mathbf{T}^N} r_\varepsilon \Phi_\varepsilon\, dxdt$ tends to zero as $\varepsilon \downarrow 0$ since

$$\left| \int_0^T \int_{\mathbf{T}^N} r_\varepsilon \Phi_\varepsilon\, dxdt \right| \le \|\Phi_\varepsilon\|_{L^\infty(0,T;L^\infty(\mathbf{T}^N))} \int_0^T \|r_\varepsilon\|_{L^1(\mathbf{T}^N)}(t)\, dt$$

and $\|\Phi_\varepsilon\|_{L^\infty}(t) \le \|\Phi\|_{L^\infty}(t)$ and $r_\varepsilon \to 0$ in $L^1\left(0, T; L^1(\mathbf{T}^N)\right)$. Similarly, the term $\int_0^T \int_{\mathbf{T}^N} u_\varepsilon \psi_\varepsilon\, dxdt$ tends to zero as $\varepsilon \downarrow 0$. Thus, sending $\varepsilon$ to zero, we deduce

$$\int_0^T \int_{\mathbf{T}^N} u\phi dxdt = 0.$$

□

## 2.4   Flow Map and Transport Equation

In this section, we shall give the uniqueness of a flow map $X \mapsto x(t, X)$ stated in Theorem 2.2 (ii) using the transport equation. The following discussion admits the existence part (Theorem 2.2 (i)).

***Proof of Theorem 2.2 (ii)*** If there are two different flow maps $x_1(t, X)$ and $x_2(t, X)$, then there is at least one $u_0 \in C^\infty(\mathbf{T}^N)$ such that $u_0(x_1(t, X)) \neq u_0(x_2(t, X))$ for some $t$ and a set of $X$ of positive measure.

For any $u_0 \in C^\infty(\mathbf{T}^N)$, we must prove the uniqueness of $u_0(x(t, X))$. Thanks to Theorem 2.4, it suffices to prove that $u(X, t) = u_0(x(t, X))$ is the unique (weak) solution of (2.7) in $L^\infty\left(0, T; L^\infty(\mathbf{T}^N)\right)$. For notational convenience in the rest of the proof, we will write the flow map $x(t, X)$ by $\varphi(t, x)$, so that the variable in (2.7) becomes $x$ rather than $X$ and $u(x, t) = u_0(\varphi(t, x))$.

For each $\psi \in C^\infty(\mathbf{T}^N)$, $h > 0$, $t \in \mathbf{R}$, we set

$$\Delta_h(t) = \int_{\mathbf{T}^N} \frac{1}{h}\{u(x, t + h) - u(x, t)\}\, \psi(x)dx$$

$$= \int_{\mathbf{T}^N} \frac{1}{h}\{u_0(\varphi(t + h, x)) - u_0(\varphi(t, x))\}\, \psi(x)dx.$$

By the group property (2.5), we see that

$$\Delta_h(t) = \int_{\mathbf{T}^N} \frac{1}{h} \{u_0 \left(\varphi \left(t, \varphi(h, x)\right)\right) - u_0 \left(\varphi(t, x)\right)\} \, \psi(x) dx.$$

By the volume-preserving property (2.2), we see that

$$\int_{\mathbf{T}^N} u_0 \left(\varphi \left(t, \varphi(h, x)\right)\right) \psi(x) dx = \int_{\mathbf{T}^N} u_0 \left(\varphi(t, z)\right) \psi \left(\varphi(-h, z)\right) dz,$$

where we take $z = \varphi(h, x)$. Thus, we observe that

$$\Delta_h(t) = \int_{\mathbf{T}^N} \frac{1}{h} u(z, t) \{\psi \left(\varphi(-h, z)\right) - \psi(z)\} \, dz. \tag{2.16}$$

We next note that $(b \circ \varphi) \cdot (\nabla \psi \circ \varphi) \in L^{\infty} \left(\mathbf{R}, L^1(\mathbf{T}^N)\right)$ since $\varphi$ has the volume-preserving property (2.2), which implies

$$\int_{\mathbf{T}^N} |b \circ \varphi(x)| \, dx = \int_{\mathbf{T}^N} |b \left(\varphi(t, x)\right)| \, dx = \int_{\mathbf{T}^N} |b(x)| \, dx,$$

$$\int_{\mathbf{T}^N} |\nabla \psi \circ \varphi(x)| \, dx = \int_{\mathbf{T}^N} |\nabla \psi \left(\varphi(t, x)\right)| \, dx = \int_{\mathbf{T}^N} |\nabla \psi(x)| \, dx.$$

Moreover,

$$\frac{\partial}{\partial t}(\psi \circ \varphi) = (b \circ \varphi) \cdot (\nabla \psi \circ \varphi)$$

since $\partial_t \varphi(t, x) = b \left(\varphi(t, x)\right)$ for a.e. $x \in \mathbf{T}^N$ and for all $t \in \mathbf{R}$ by (i). Thus,

$$\psi \left(\varphi(-h, z)\right) - \psi(z) = - \int_0^h b \left(\varphi(-\sigma, z)\right) \cdot (\nabla \psi) \left(\varphi(-\sigma, z)\right) d\sigma.$$

We plug this formula into (2.16) to get

$$\Delta_h(t) = - \int_{\mathbf{T}^N} \frac{1}{h} u(z, t) \left[ \int_0^h b \left(\varphi(-\sigma, z)\right) \cdot (\nabla \psi) \left(\varphi(-\sigma, z)\right) d\sigma \right] dz.$$

As previously, we invoke the volume-preserving property (2.2) and the group property (2.5) to get

$$\Delta_h(t) = - \int_{\mathbf{T}^N} \frac{1}{h} b(x) \cdot \nabla \psi(x) \int_0^h u \left(\varphi(\sigma, x), t\right) d\sigma dx$$

$$= - \int_{\mathbf{T}^N} b(x) \cdot \nabla \psi(x) \frac{1}{h} \int_0^h u_0 \left(\varphi \left(t, \varphi(\sigma, x)\right)\right) d\sigma dx$$

$$= - \int_{\mathbf{T}^N} b(x) \cdot \nabla \psi(x) \frac{1}{h} \int_0^h u(x, t + \sigma) d\sigma dx.$$

Since $\varphi = \varphi(t, x)$ is continuous in $t \in \mathbf{R}$ for a.e. $x \in \mathbf{T}^N$,

$$\lim_{h\downarrow 0} \frac{1}{h} \int_0^h u(x, t + \sigma)\, d\sigma = u(x, t) \quad \text{for all} \quad t$$

for a.e. $x \in \mathbf{T}^N$. Since $b \cdot \nabla\psi \in L^1\left(\mathbf{T}^N \times (0, T)\right)$ and $u$ is bounded on $\mathbf{T}^N \times (0, T)$ independent of $h$, by the dominated convergence theorem we conclude that

$$\Delta_h(t) \to -\int_{\mathbf{T}^N} b(x) \cdot (\nabla\psi)(x)u(x, t)\, dx \quad \text{as} \quad h \downarrow 0; \tag{2.17}$$

this convergence is *locally uniform* in $(0, T)$, i.e.,

$$\lim_{h\downarrow 0} \sup_{a \le t \le b} |\Delta_h(t) - \Psi(t)| = 0$$

for any $[a, b] \subset (0, T)$, where $\Psi$ denotes the right-hand side of (2.17).

It is easy to see that

$$\Delta_h(t) \to \frac{\partial}{\partial t} \int_{\mathbf{T}^N} u(x, t)\psi(x) dx \quad \text{as} \quad h \downarrow 0$$

in the sense of distribution as a function of $t$. We thus conclude that $u$ satisfies (2.8) for $\phi(x, t) = \psi(x)\eta(t)$, with $\eta \in C_c^\infty([0, T))$. Thus, (2.8) is still valid for any $\phi \in C_c^\infty\left(\mathbf{T}^N \times [0, T)\right)$ since the linear span of the product type is dense in the class of test functions $C_c^\infty\left(\mathbf{T}^N \times [0, T)\right)$. $\qquad\square$

▶ **Remark 2.10** In the case of $\mathbf{T}^N$, (2.3) for general $\beta$ is not invoked for the uniqueness proof; we only use $\beta = $ identity. However, if one considers the problem in $\mathbf{R}^N$ instead of $\mathbf{T}^N$, it is important to approximate the identity since, in general, only bounded $\beta$ with bounded $|b(z)| / (1 + |z|)$ is allowed. This restriction is important to understand (2.3) in the distribution sense.

## 2.5 Notes and Comments

### Remarks on Flow Maps and Transport Equations
The contents of Chap. 2 is an active area of current research. The construction of such a flow map $x = x(t, X)$ for non-Lipschitz vector field $b$ is extended when $b$ is just in $BV$ spaces [2]. Although the flow map $x = x(t, X)$ is defined only for almost all $X \in \mathbf{T}^N$, it is known that $x$ is Lipschitz in $X$ with a small exceptional set [4]. The estimate is now quantified by [30]. It is of the following form. For given $T > 0$, $p > 1$, and small $\varepsilon > 0$ there is a compact set $K$ such that $\mathcal{L}^N(\mathbf{T}^N \setminus K) < \varepsilon$ and

$$|x(t, X_1) - x(t, X_2)| \le \exp\left(C_N A_p(x)/\varepsilon^{1/p}\right) |X_1 - X_2|, \quad X_1, X_2 \in K, \ t \in [0, T],$$

with $C_N$ depending only on the dimension. Here,

$$A_p(x) = \left\{ \int_{\mathbf{T}^N} \left( \sup_{0 \le t \le T} \sup_{0 < r < 2} \frac{1}{\mathcal{L}^N (B_r(X))} \right. \right.$$

$$\left. \left. \int_{B_r(X)} \log \left( \frac{|x(t, X) - x(t, Y)|}{r} + 1 \right) dY \right)^p dX \right\}^{1/p}.$$

For simplicity, we assume that $B_2(X)$ covers the fundamental domain $\Omega$ of $\mathbf{T}^N$. The quantity $A_p(x)$ is uniformly controlled by $\|Db\|_{L^1}$ provided that $\operatorname{div} b = 0$. In [30], $\operatorname{div} b$ may not be zero, but some uniform compressibility for the flow map is assumed. Moreover, in [30], the flow map itself is studied directly without using the transport equations.

Our strategy for proving the uniqueness of the flow map in Chap. 2 is to reduce the uniqueness of the transport equation, as stated in Theorem 2.4. However, we warn the reader that the uniqueness of the transport equation fails if one considers a less regular vector field. In fact, if one relaxes the assumption

$$b \in W^{1,p'}(\mathbf{T}^N), \quad \operatorname{div} b = 0$$

by

$$b \in L^{p'}(\mathbf{T}^N), \quad Db \in L^{\tilde{p}}(\mathbf{T}^N), \quad \operatorname{div} b = 0,$$

with $1/p + 1/\tilde{p} > 1 - 1/(N - 1)$, then the assertion of Theorem 2.4 fails. In other words, there is a nontrivial weak solution $u$ to (2.7) with zero initial data. This is first proved by Modena and Székelyhidi, Jr. [75] using a convex integration method. A solution constructed there is not a renormalized solution, i.e., the assertion of Lemma 2.5 does not hold for their solution $u$. This can be understood as meaning there is a microscopic effect that cannot be captured by the macroscopic notion of a weak solution. Recently, a nonrenormalized weak solution was constructed by Drivas et al. [34] using a vanishing viscosity method with anomalous dissipation. As pointed out in [86], such a solution is produced by a microscopic effect. The notion of a weak solution is too weak to guarantee uniqueness even for linear transport equations. In a very recent preprint, Huysmans and Titi [55] proved that the uniqueness may fail even among renormalized solutions if one only assumes that $b = b(x, t)$ is bounded with $\operatorname{div} b = 0$. (Note that their $b$ depends on time $t$.) They constructed two different solutions which are given as subsequential vanishing viscosity limits, of the same equation.

In the next two chapters, we will discuss scalar conservation laws and the Hamilton–Jacobi equations, where a naive "weak solution" may not be unique. For these equations one is able to recover uniqueness by considering a special class of weak solutions.

It is of current interest to show the nonuniqueness of weak solutions for various physically important nonlinear equations, even if the viscosity is included, for example, the Navier–Stokes equations [20]. However, it is not clear what kind of extra condition would guarantee uniqueness.

The contents of Chap. 2 are taken from the paper [32], where the problem is studied on $\mathbf{R}^N$. In this book, we consider the problem on $\mathbf{T}^N$ to simplify the situation. Lemma 2.7 is a crucial step of the argument and is often called DiPerna–Lion's lemma. A variant of this lemma is called Friedrichs' commutator lemma in [39, Section 11.19]. This type of lemma is useful for studying mass conservation laws for compressible flows.

## 2.6 Exercises

2.1 Give an example of the nonuniqueness of a solution to (1.1) with a given initial datum when $b \in \bigcap_{p \geq 1} W^{1,p}(\mathbf{R}^N)$.

2.2 Set

$$L = \{\phi : \mathbf{T}^N \to \mathbf{R} \mid \phi \text{ is Lebesgue measurable}$$

$$\text{and } |\phi| < \infty \text{ a.e.}\}.$$

Set $d(\phi, \psi) = \|\min(|\phi - \psi|, 1)\|_{L^1(\mathbf{T}^N)}$. Show that $(L, d)$ is a metric space.

2.3 Let $f$ be a locally integrable function in $(0, T)$. Assume that

$$\int_0^T f(t)\psi(t)\,dt = 0$$

for all $\psi \in C_c^\infty((0, T))$. Show that $f(t) = 0$ for almost all $t \in (0, T)$.

2.4 Let $\rho_\varepsilon$ be a Friedrichs' mollifier. Let $f$ be continuous on $\mathbf{R}$. Show that $\rho_\varepsilon * f$ is in $C^\infty(\mathbf{R})$.

2.5 In the context of Exercise 2.4, show that $\rho_\varepsilon * f$ converges to $f$ locally uniformly in $\mathbf{R}$ as $\varepsilon$ tends to zero.

See [45] for details of Exercises 2.3–2.5.

2.6 Let $L$ be the space defined in Exercise 2.2. Set

$$\overline{d}(\phi, \psi) = \int_{\mathbf{T}^N} \frac{|\phi(x) - \psi(x)|}{1 + |\phi(x) - \psi(x)|}\,dx.$$

Show that $(L, \overline{d})$ is a metric space.

2.7 Assume that $\{f_m\}_{m=1}^\infty$ is a sequence converging to $f$ in $L^1(\mathbf{T}^N)$ as $m \to \infty$. In other words,

$$\lim_{m \to \infty} \int_{\mathbf{T}^N} |f_m(x) - f(x)|\,dx = 0.$$

Show that $\{f_m\}$ converges to $f$ in measure.

2.8 Assume that $\varphi : \mathbf{T}^N \to \mathbf{T}^N$ is a volume-preserving mapping. In other words, $\varphi$ has the property that

$$\mathcal{L}^N \left( \left\{ x \in \mathbf{T}^N \mid \varphi(x) \in A \right\} \right) = \mathcal{L}^N(A)$$

for any measurable set $A$. Show that

$$\int_{\mathbf{T}^N} \psi\left( \varphi(x) \right) dx = \int_{\mathbf{T}^N} \psi(x) \, dx$$

for any measurable function $\psi$ on $\mathbf{T}^N$.

2.9 (i) For $p \in [1, \infty)$, let $p'$ denote the conjugate exponent of $p$, i.e., $1/p + 1/p' = 1$. Assume that a sequence $\{f_m\}_{m=1}^\infty$ converges to $f$ in $L^p(\mathbf{T}^N)$ as $m \to \infty$. In other words,

$$\lim_{m \to \infty} \int_{\mathbf{T}^N} |f_m(x) - f(x)|^p \, dx = 0.$$

Assume that a sequence $\{g_m\}_{m=1}^\infty$ converges *-weakly to $g$ in $L^{p'}(\mathbf{T}^N)$ as $m \to \infty$. In other words,

$$\lim_{m \to \infty} \int_{\mathbf{T}^N} g_m(x)\varphi(x) \, dx = \int_{\mathbf{T}^N} g(x)\varphi(x) \, dx$$

holds for all $\varphi \in L^p(\mathbf{T}^N)$. Show that

$$\lim_{m \to \infty} \int_{\mathbf{T}^N} f_m(x)g_m(x) \, dx = \int_{\mathbf{T}^N} f(x)g(x) \, dx.$$

(ii) Set $f_m(x) = \sin mx \in L^2(\mathbf{T})$, where $\mathbf{T} = \mathbf{R}/(2\pi \mathbf{Z})$. Show that $\{f_m\}_{m=1}^\infty$ converges weakly to 0 in $L^2(\mathbf{T})$ but

$$\lim_{m \to \infty} \int_{\mathbf{T}^N} f_m(x)^2 \, dx \neq 0.$$

2.10 Let $\rho_\varepsilon$ be a Friedrichs' mollifier. For $f \in L^\infty(\mathbf{T}^N)$, show that $\rho_\varepsilon * f$ converges to $f$ a.e. as $\varepsilon$ tends to zero. Moreover, show that

$$\|f\|_{L^\infty(\mathbf{T}^N)} = \lim_{\varepsilon \downarrow 0} \|f_\varepsilon\|_{L^\infty(\mathbf{T}^N)}, \quad \|f_\varepsilon\|_{L^\infty(\mathbf{T}^N)} \leq \|f\|_{L^\infty(\mathbf{T}^N)}.$$

# Uniqueness of Solutions to Initial Value Problems for a Scalar Conversation Law

In Chap. 2, we discussed the uniqueness of a weak solution to a transport equation, which is linear and of the first order. In this chapter, we consider scalar conservation laws, which are quasilinear but still of the first order. The major difference between the linear transport equations with a divergence-free (solenoidal) coefficient and a conservation law lies in the uniqueness problem of a weak solution. For the transport equation, it is unique under a very weak regularity assumption. However, for a conversation law, it may not be unique under a reasonable regularity assumption allowing discontinuities. To recover uniqueness, one must introduce an extra condition, called an entropy condition, that is not a regularity condition. Another difference is that the solution may develop singularity even if the initial datum are smooth for a conservation law but the solution is smooth for the transport equation if all data and coefficients are smooth.

In this chapter, we introduce a scalar conservation law and observe that a discontinuity –called a shock– may develop in finite time. To track the whole evolution, we need to introduce a weak solution. However, unfortunately, weak solutions may not be unique. To recover uniqueness, we introduce the "entropy condition" and the notion of an "entropy solution." After discussing the entropy condition, we prove the uniqueness of an entropy solution. To avoid technical complications, we discuss uniqueness in a periodic setting. A key idea in proving uniqueness is a method of doubling variables that is due to Kružkov [68]. The contents of this chapter are essentially taken from a book [53] by Holden and Risebro, with the modification that the uniqueness is discussed in a periodic setting. This topic is also discussed in [36, Chapter 11], with an emphasis on systems of conservation laws.

© The Author(s), under exclusive license to Springer Nature Switzerland AG 2023
M.-H. Giga, Y. Giga, *A Basic Guide to Uniqueness Problems for Evolutionary Differential Equations*, Compact Textbooks in Mathematics,
https://doi.org/10.1007/978-3-031-34796-2_3

## 3.1    Entropy Condition

In this section, we introduce a scalar conservation law and discuss the discontinuity of a solution. If initial datum are smooth, we are able to solve the equation locally in time, but it may develop discontinuity. To track evolution globally in time, we introduce the notion of a weak solution by integration by parts. We notice that uniqueness may be violated. There are several types of discontinuity. We only allow a particular type of discontinuity that satisfies the entropy condition. This eventually leads to the notion of an entropy solution.

### 3.1.1   Examples

We consider a flow map $x(t, X)$ generated by a vector field $u$ on $\mathbf{R}^N$, i.e.,

$$\dot{x}(t, X) = u\,(x(t, X), t) \quad \text{for} \quad t > 0, \quad x(0, X) = X,$$

where $\dot{x}(t, X) = \frac{\partial}{\partial t} x(t, X)$. The coordinate by $X$ is often called the Lagrangian coordinate, while the coordinate by $x$ is called the Euler coordinate.

Assume that there is no acceleration. Physically speaking, there is no force by Newton's law. Then

$$\ddot{x}(t, X) = 0 \quad \text{or} \quad \frac{\partial^2}{\partial t^2} x(t, X) = 0,$$

where the partial derivative is taken in the Lagrangian coordinate. We shall write this law for $u(x, t)$ for the Euler coordinate. Since

$$\ddot{x} = \nabla_x u \cdot \dot{x} + u_t \quad \text{with} \quad \dot{x} = u(x, t) \quad \text{or}$$

$$\ddot{x}^i = \sum_{j=1}^{N} \partial_{x_j} u^i \dot{x}^i + u_t^i \quad \text{with} \quad \dot{x}^i = u^i(x, t),$$

where the partial derivative in the direction of $x, t$ of $u$ is in the Euler coordinate, we see that $\ddot{x} = 0$ is equivalent to saying that

$$u_t + u \cdot \nabla_x u = 0 \quad \text{or} \quad u_t^i + \sum_{j=1}^{N} u^j \partial_{x_j} u^i = 0, \quad 1 \leq i \leq N.$$

If $N = 1$, this is simply

$$u_t + u\,u_x = 0 \quad \text{or} \quad u_t + \left(\frac{u^2}{2}\right)_x = 0, \tag{3.1}$$

which is called the Burgers equation. Here $u_x = \partial u / \partial x$. This equation is a typical example of a (scalar) conservation law

$$u_t + f(u)_x = 0, \tag{3.2}$$

where $f$ is a function of $u$ and $f(u)_x = \frac{\partial}{\partial x}(f(u)) = \frac{\partial}{\partial x}(f \circ u)(x)$. In (3.1), $f(u) = u^2/2$.

We give another derivation of a conservation law modeling a traffic flow. We consider the simplest situation: a road having only one lane parameterized by a single coordinate $x$. All cars are assumed to move in only one direction, that of increasing $x$. Let $\rho(x, t)$ be the (number) density of cars at location $x$ and time $t$. The number of cars in the interval $[a, b]$ at time $t$ corresponds to $\int_a^b \rho(x, t) dx$. Let $v(x, t)$ be the velocity of the car at $x$. The rate of cars passing a point $x$ at some time $t$ is given by $v(x, t)\rho(x, t)$. Thus, the change ratio of the number of cars in $[a, b]$ should be

$$\frac{d}{dt} \int_a^b \rho(x, t) dx = -(v(b, t)\rho(b, t) - v(a, t)\rho(a, t)).$$

Since the right-hand side equals $-\int_a^b (v\rho)_x dx$ and since $(a, b)$ is arbitrary, we get

$$\rho_t + (\rho v)_x = 0, \tag{3.3}$$

which is a typical mass conservation law, for example, in fluid mechanics. (In a multidimensional setting, it must be that

$$\rho_t + \text{div}(\rho v) = 0,$$

which is the fundamental mass conservation law in science. Here $v$ is a vector field.) In the simplest model, the velocity $v$ is assumed to be a given function of the (number) density $\rho$ only. This one-dimensional model may approximate the situation where the road is uniform with no obstacles like signals, crossings, or curves forcing cars to slow down. We postulate that there is a uniform maximal speed $v_{\max}$ for any car. If traffic is light, a car will approach this maximal speed, but the car will have to slow down if the number of cars increases. If $\rho$ reaches some value $\rho_{\max}$, all cars must stop. Thus, it is reasonable to assume that $v$ is a monotone decreasing function of $\rho$ such that $v(0) = v_{\max}(> 0)$, $v(\rho_{\max}) = 0$. The simplest function is a linear function, i.e.,

$$v(\rho) = v_{\max}(1 - \rho/\rho_{\max}) \quad \text{for} \quad \rho \in [0, \rho_{\max}] \tag{3.4}$$

**Fig. 3.1** Profile of $V$

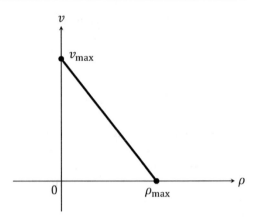

(Figure 3.1). If $\tilde{u} = \rho/\rho_{max}$, $\tilde{x} = v_{max}x$ is normalized, the resulting normalized equation of (3.3) with (3.4) for $\tilde{u} = \tilde{u}(\tilde{x}, t)$ is of the form

$$\tilde{u}_t + (\tilde{u}(1 - \tilde{u}))_{\tilde{x}} = 0 \quad \text{for} \quad \tilde{u} \in [0, 1].$$

For further reference, we rewrite this equation as

$$u_t + (u(1 - u))_x = 0 \tag{3.5}$$

by writing $u = \tilde{u}$, $x = \tilde{x}$. The Burgers equation (3.1) is obtained by setting $\tilde{u} = \frac{1}{2}(1 - u)$, $\tilde{x} = x$.

### 3.1.2   Formation of Singularities and a Weak Solution

An important feature of conservation law (3.1) is that the solution may become singular in finite time.

---

**Proposition 3.1**

*Assume that $f$ is smooth in $\mathbf{R}$ and that its second derivative $f''$ is positive in an interval $[\alpha, \beta]$, which is nontrivial, i.e., $\alpha < \beta$. Let $u_0 \in C^\infty(\mathbf{R})$ be nonincreasing and $u_0(x) = \beta$ for $x < -x_0$ and $u_0(x) = \alpha$ for $x > x_0$ with some $x_0 > 0$. Then there exists a unique smooth solution $u$ of (3.2), with $u(0, x) = u_0(x)$, for $x \in \mathbf{R}$ satisfying $\alpha \leq u \leq \beta$ in $\mathbf{R} \times (-T_0, T_1)$, with some $T_0, T_1 > 0$, but the maximal (forward) existence time $T_1$ must be finite.*

**Fig. 3.2** Graph of $u_0$

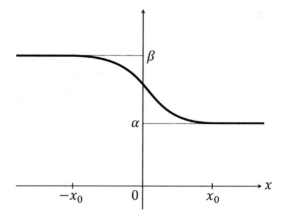

*Proof.* We consider the equation for $v \in \mathbf{R}$ of the form

$$v = u_0 \left( x - f'(v)t \right) \tag{3.6}$$

for a given $x, t \in \mathbf{R}$. Here, $f'$ denotes the derivative of $f$ when $f$ depends on just one variable. See Fig. 3.2 for the profile of $u_0$. This equation has a unique solution $\bar{v} \in [\alpha, \beta]$ for all $x \in \mathbf{R}$ provided that $t$ is sufficiently small, say, $|t| < t_0$, with some $t_0 > 0$ by the implicit function theorem [67]. Indeed, differentiating

$$F(v, x, t) = v - u_0 \left( x - f'(v)t \right)$$

with respect to $v$ we get

$$\frac{\partial F}{\partial v}(v, x, t) = 1 + u_0' \left( x - f'(v)t \right) f''(v)t.$$

This is bounded away from zero uniformly in $x$ and small $t$, allowing negative $t$, say, $|t| < t_0$ since $f''$ is bounded in $[\alpha, \beta]$ and $u_0'$ is bounded. Then, by the implicit function theorem, we get a unique $v = \bar{v}$, solving (3.6).

We shall write $\bar{v} = u(x, t)$ since $\bar{v}$ depends on $(x, t)$. Since $\bar{v}$ solves (3.6), we see that $F(u(x, t), x, t) = 0$ for $x \in \mathbf{R}$, $t$, with $|t| < t_0$. Since $F$ depends on $v, x$ and $t$ smoothly, we conclude that $u$ is smooth in $\mathbf{R} \times (-t_0, t_0)$ by the smooth dependence of parameters in the implicit function theorem. (The curve $z = x - f'(u_0(z))t$ in the $xt$-plane with a parameter $z \in \mathbf{R}$ is often called a *characteristic curve* (Fig. 3.3). The value of $u$ on each characteristic curve $z = x - f'(u_0(z))t$ equals the constant $u_0(z)$ by (3.6). Unlike the linear equation (2.6), the characteristic curve may depend on the initial datum $u_0$.)

**Fig. 3.3** Characteristic curves

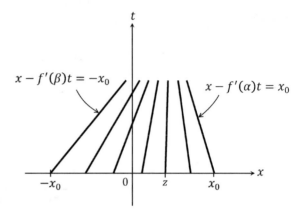

Differentiating both sides of (3.6) by setting $v = u(x, t)$, we get

$$u_t = u_0' \left( x - f'(u)t \right) \left( -f''(u)u_t t - f'(u) \right),$$
$$f'(u)u_x = u_0' \left( x - f'(u)t \right) \left( -f''(u)u_x t + 1 \right) f'(u).$$

Adding both sides we get

$$u_t + f'(u)u_x = u_0' \left( x - f'(u)t \right) \left( -f''(u)t \left( u_t + f'(u)u_x \right) \right).$$

From this identity we see that $u$ solves (3.2) in $\mathbf{R} \times (-t_0, t_1)$, with $u(x, 0) = u_0(x)$, $x \in \mathbf{R}$, if we choose a sufficiently small $t_1 \in (0, t_0)$. Indeed, this identity implies $u_t + f'(u)u_x = 0$ unless $u_0' \left( u - f'(u)t \right) \left( -f''(u)t \right) = 1$. However, the last identity does not hold for $t < 0$ since $u_0' \leq 0$ and $f''(u) > 0$, and also for small $t > 0$ independent of $x$ since $u_0'$ and $f'' \left( u(x, t) \right)$ are bounded. Thus, we get (3.2).

The uniqueness can be proved easily since the difference $w := u_1 - u_2$ of two solutions $u_1$ and $u_2$ solves

$$w_t + (pw)_x = 0, \quad w|_{t=0} = 0,$$

with

$$p(x, t) = \int_0^1 f' \left( u_2 + \theta(u_1 - u_2) \right) d\theta,$$

which is smooth and bounded with its derivatives. Indeed,

$$f(u_1) - f(u_2) = \int_0^1 \frac{d}{d\theta} \left( f \left( u_2 + \theta(u_1 - u_2) \right) \right) d\theta = pw$$

so we get the preceding $w$ equation by subtracting equation (3.2) for $u_2$ from that for $u_1$. We next apply an idea of the method of characteristics (see Chap. 2, especially the paragraph including (2.6)) to this $w$ equation

$$w_t + p w_x + p_x w = 0.$$

In general, it is more involved since $p$ depends on time $t$. Here we simply use it as a change of variables to remove the $w_x$ term. Let $x = x(t, X)$ be the unique solution of

$$\dot{x} = p(x, t) \text{ for small } |t|, \quad x(0) = X.$$

We set

$$W(X, t) := w(x(t, X), t)$$

and observe that

$$\frac{\partial W}{\partial t} = w_t + p w_x.$$

The $w$ equation is transformed to

$$W_t + q W = 0, \quad W|_{t=0} = 0$$

for small $|t|$, where $q = p_x(x(t, X), t)$. This is a linear ordinary differential equation, so the uniqueness (Proposition 1.1) yields $W \equiv 0$. Thus, $w \equiv 0$ on $\mathbf{R} \times (-\delta, \delta)$ for small $\delta > 0$. A similar argument implies that the time interval $[t_-, t_+]$ where uniqueness $w = 0$ holds is open. Thus, $w \equiv 0$ on $(-t_0, t_1)$, i.e., $u_1 \equiv u_2$ on $\mathbf{R} \times (-t_0, t_1)$.

By (3.6) we see that

$$u(x, t) = \beta \quad \text{for} \quad x - f'(\beta)t < -x_0,$$
$$u(x, t) = \alpha \quad \text{for} \quad x - f'(\alpha)t > x_0.$$

Since $\alpha < \beta$, for sufficiently large $t$ the two characteristic curves $x_0 = x - f'(\alpha)t$ and $-x_0 = x - f'(\beta)t$ merge (Fig. 3.3). Let $t = t_*$ be a number such that $f'(\alpha)t_* + x_0 < f'(\beta)t_* - x_0$. Then $u(\cdot, t_*)$ has two values, $\alpha$ and $\beta$, on $\left(f'(\alpha)t_* + x_0, f'(\beta)t_* - x_0\right)$. Thus, $t_1 < t_*$. This implies that the (forward) maximal existence time for a smooth solution is finite.                       □

We shall consider the initial value problem to (3.2) for $t > 0$. By Proposition 3.1, we must introduce a notion of a weak solution as in Definition 2.3 to track the whole evolution of a solution.

### Definition 3.2

Assume that $f \in C(\mathbf{R})$. For $u_0 \in L^\infty(\mathbf{R})$, we say that $u \in L^\infty(\mathbf{R} \times (0, T))$ is a *weak solution* of (3.2) with initial datum $u_0$ if

$$\int_{\mathbf{R} \times (0,T)} \{\varphi_t u + \varphi_x f(u)\} \, dx dt + \int_{\mathbf{R}} \varphi|_{t=0} u_0 dx = 0 \tag{3.7}$$

for all $\varphi \in C_c^\infty(\mathbf{R} \times [0, T))$. If $u_0$ and $u$ is periodic in $x$, i.e., a function on $\mathbf{T} = \mathbf{R}/\omega_1 \mathbf{Z}$ with some $\omega_1 > 0$, then $\varphi$ should be taken from $C_c^\infty(\mathbf{T} \times [0, T))$.

We shall discuss the speed of jump discontinuity. Its speed is represented by the magnitude of the jump, and such a representation is called the *Rankine–Hugoniot condition*. Let $x(t)$ be a $C^1$ function defined on an interval $[t_0, t_1]$, with $t_0 < t_1$, $t_0, t_1 \in \mathbf{R}$. Let $D = J \times (t_0, t_1)$ be an open set containing the graph of $x(t)$ in $(t_0, t_1)$, where $J$ is an open interval in $\mathbf{R}$. We set

$$D_r = \{(x, t) \in D \mid x > x(t)\},$$

$$D_\ell = \{(x, t) \in D \mid x < x(t)\},$$

$$\Gamma = \overline{D_r} \cap \overline{D_\ell}.$$

Here, $\Gamma$ is simply the graph of the curve $x = x(t)$. See Fig. 3.4.

**Fig. 3.4** Sets $D_\ell$, $D_r$ and $\Gamma$

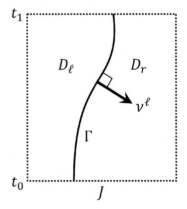

---

**Lemma 3.3**
*Let $f \in C(\mathbf{R})$ be given. Let $u$ be $C^1$ in $\overline{D_r}$ and $\overline{D_\ell}$, and let $u$ satisfy (3.7) for all $\varphi \in C_c^\infty (D \times (t_0, t_1))$. Then*

$$\dot{x}(t)(u_\ell - u_r) = f_\ell - f_r \tag{3.8}$$

*for $t \in (t_0, t_1)$, with*

$$u_\ell = \lim \{u(y, s) \mid (y, s) \to (x(t), t), (y, s) \in D_\ell\} \text{ (left limit)},$$

$$u_r = \lim \{u(y, s) \mid (y, s) \to (x(t), t), (y, s) \in D_r\} \text{ (right limit)},$$

*and $f_\ell = f(u_\ell)$, $f_r = f(u_r)$. (The speed $s = \dot{x}(t)$ is called the speed of the shock.) Conversely, if $u$ satisfies (3.2) in $D_r$ and $D_\ell$ and satisfies (3.8), then $u$ satisfies (3.7) for all $\varphi \in C_c^\infty (D \times (t_0, t_1))$.*

---

*Proof.* Since $u$ is a classical solution of (3.2) in each $D_i$ ($i = r, \ell$), integration by parts yields

$$\int_{D_i} \{\varphi_t u + \varphi_x f(u)\} \, dx dt = \int_{\partial D_i} (v_t u + v_x f(u)) \, \varphi d\mathcal{H}^1$$

$$= \int_\Gamma \left(v_t^i u_i + v_x^i f_i\right) \varphi d\mathcal{H}^1,$$

where $(v_x^i, v_t^i)$ is an external unit normal of $\partial D_i$. Here, $d\mathcal{H}^1$ denotes the line element of the curve $x = x(t)$. Since $u$ is a "weak solution" of (3.2) in $D$ (i.e., $u$ satisfies (3.7) for all $\varphi \in C_c^\infty (D \times (t_0, t_1))$), we see that

$$\int_\Gamma \left\{\left(v_t^r u_r + v_t^\ell u_\ell\right) + \left(v_x^r f_r + v_x^\ell f_\ell\right)\right\} \varphi d\mathcal{H}^1 = 0.$$

Since $v^r = -v^\ell$ and $\varphi$ is arbitrary, we now conclude (cf. Exercise 2.3) that

$$v_t^\ell (u_\ell - u_r) + v_x^\ell (f_\ell - f_r) = 0.$$

Since

$$\left(v_x^\ell, v_t^\ell\right) = (1, -\dot{x}(t)) / \left(1 + (\dot{x}(t))^2\right)^{1/2},$$

the desired relation (3.8) follows. Checking this argument carefully, we see the converse is easily obtained. The relation (3.8) is called the *Rankine–Hugoniot condition*. □

### 3.1.3   Riemann Problem

We consider the following special initial value problem for (3.2), which is called the *Riemann problem*. The initial datum we consider are

$$u_0(x) = \begin{cases} u_\ell, & x < 0, \\ u_r, & x > 0, \end{cases} \tag{3.9}$$

where $u_\ell$ and $u_r$ are constants, i.e., $u_\ell, u_r \in \mathbf{R}$.

For simplicity, we assume that $u_\ell > u_r$ in this subsection. It is easy to see that

$$u_S(x,t) = \begin{cases} u_\ell, & x < x(t), \\ u_r, & x > x(t) \end{cases} \tag{3.10}$$

is a weak solution of (3.2) with (3.9) provided that $x(t) = t(f_\ell - f_r)/(u_\ell - u_r)$ by (3.8). If $u_r < u_\ell$ and $f$ is convex, it turns out that this is the only weak solution. However, in the case where $u_r < u_\ell$ and $f$ is concave, there is another weak solution called a *rarefaction wave*. Instead of writing a general form of a solution, we just restrict ourselves to the traffic flow equation (3.5) where $f(u) = u(1 - u)$. In this case, the function

$$u_R(x,t) = \begin{cases} u_\ell, & x < x_\ell(t), \\ \dfrac{1}{2} - \dfrac{x}{2t}, & x_\ell(t) \le x \le x_r(t), \\ u_r, & x > x_r(t), \end{cases} \tag{3.11}$$

with $x_\ell(t) = \left(\frac{1}{2} - u_\ell\right) 2t$, $x_r(t) = \left(\frac{1}{2} - u_r\right) 2t$, is a weak solution of (3.2) with (3.9) provided that $u_r < u_\ell$ (Figs. 3.5 and 3.6). This is easy to check since there is no jump and $1/2 - x/(2t)$ solves equation (3.2) in the region $x_\ell < x < x_r$. The question is which is reasonable as a "solution." Of course, it depends on the physics we consider. For the traffic flow problem, consider the case where $u_r = 0$ and $u_\ell = 1$. The solution $u_S$ in this case is time-independent since $f_\ell(0) = f_\ell(1) = 0$,

**Fig. 3.5**   A rarefaction wave
$u_R$ at time $t$

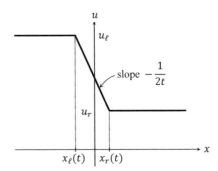

**Fig. 3.6** Characteristic
curves

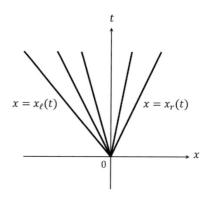

so that $x(t) = 0$. Is it natural to stop even if there are no cars in front of us? There is
no signal. From our intuition, $u_R$ looks like a more reasonable solution. The question
is how we determine this.

### 3.1.4  Entropy Condition on Shocks

We consider the viscous regularization of (3.2) of the form

$$u_t^\varepsilon + f(u^\varepsilon)_x = \varepsilon u_{xx}^\varepsilon \tag{3.12}$$

with initial datum $u_0$ of the form of (3.9). We are interested in the case where the
limit tends to $u_S$ as $\varepsilon \to 0$. We seek the solution $u^\varepsilon$ of the form

$$u^\varepsilon(x, t) = U\left(\frac{x - st}{\varepsilon}\right), \tag{3.13}$$

where $s$ is the shock wave speed $s = (f_\ell - f_r)/(u_\ell - u_r)$ determined by the Rankine–
Hugoniot condition. The function $U = U(\xi)$ in (3.13) must satisfy

$$-sU_\xi + (f(U))_\xi = U_{\xi\xi}$$

if $u^\varepsilon$ solves (3.12), where $U_\xi = (d/d\xi)U(\xi)$. Integrate both sides to get

$$U_\xi = -sU + f(U) + C_0, \tag{3.14}$$

where $C_0$ is a constant of integration. (We consider this equation assuming that $f$ is
$C^1$, so that its initial value problem admits only one $C^1$ solution (Proposition 1.1).
If $U$ is $C^1$, then the right-hand side of (3.14) is $C^1$, so that $U$ is $C^2$.) We postulate

that $u_S$ is the limit of $u^\varepsilon$ as $\varepsilon \downarrow 0$; then we should have

$$u_S(x, t) = \lim_{\varepsilon \to 0} U\left((x - st)/\varepsilon\right) = \begin{cases} u_\ell \text{ for } & x < st, \\ u_r \text{ for } & x > st, \end{cases}$$

so

$$\lim_{\xi \to -\infty} U(\xi) = u_\ell, \quad \lim_{\xi \to \infty} U(\xi) = u_r.$$

If we postulate $U$ is monotone, we have

$$\lim_{\xi \to \pm\infty} U_\xi(\xi) = 0$$

since $\lim_{\xi \to \pm\infty} U_\xi$ always exists by (3.14). (The monotonicity follows from the maximum principle for the derivative of $u^\varepsilon$.) Letting $\xi \to \pm\infty$ in (3.14), we obtain

$$C_0 = su_\ell - f_\ell = su_r - f_r.$$

The last equality also gives the Rankine–Hugoniot condition. Thus, we obtain an ordinary differential equation for $U$ with boundary condition at $\pm\infty$ of the form

$$\frac{\mathrm{d}}{\mathrm{d}\xi} U(\xi) = -s\left(U(\xi) - u_\ell\right), +\left(f\left(U(\xi)\right) - f_\ell\right), \tag{3.15}$$

$$U(\infty) = u_r, \quad U(-\infty) = u_\ell.$$

---

**Definition 3.4**

If there exists a solution $U$ of (3.15) with $U(\infty) = u_r$, $U(-\infty) = u_\ell$ ($u_r \neq u_\ell$), we say that $u_S$ in (3.10) with $x(t) = st$, $s = (f_\ell - f_r)/(u_\ell - u_r)$ satisfies a *traveling wave entropy condition*.

We shall derive an equivalent condition for $u_r$ and $u_\ell$, so that $u_S$ satisfies a traveling wave entropy condition.

---

**Proposition 3.5**

Let $f \in C^1(\mathbf{R})$. Assume $u_\ell < u_r$ (resp. $u_r < u_\ell$). Let $u_S$ be of the form of (3.10), with $x = st$, $s = (f_\ell - f_r)/(u_\ell - u_r)$, where $f_\ell = f(u_\ell)$ and $f_r = f(u_r)$. Then $u_S$ fulfills the traveling wave entropy condition if and only

(continued)

> **Proposition 3.5** (continued)
> *if the graph of $f(u)$ lies above (resp. below) the straight line segment joining $(u_\ell, f_\ell)$ and $(u_r, f_r)$, i.e.,*
>
> $$f(u) > f_\ell + s(u - u_\ell) = f_r + s(u - u_r)$$
> $$(resp. \ f(u) < f_\ell + s(u - u_\ell) = f_r + s(u - u_r))$$
>
> *for all $u \in (u_\ell, u_r)$ (resp. $u \in (u_r, u_\ell)$).*

*Proof.* Assume that $u_\ell < u_r$. We first observe that $U_\xi$ does not vanish. Indeed, if there were $\xi_0$ such that $U_\xi(\xi_0) = 0$, then $a = U(\xi_0)$ should satisfy

$$-s(a - u_\ell) + (f(a) - f_\ell) = 0.$$

Thus, $U \equiv a$ is a solution to (3.15), which is unique by Proposition 1.1. Thus, $U$ must be a constant that cannot achieve at least one of the boundary conditions $U(\infty) = u_r$, $U(-\infty) = u_\ell$. Thus, $U_\xi(\xi) > 0$ for all $\xi$. This implies

$$f_\ell + s(u - u_\ell) < f(u)$$

for $u \in (u_\ell, u_r)$. Recalling the Rankine–Hugoniot condition, $s = (f_\ell - f_r)/(u_\ell - u_r)$, we observe the desired condition (Fig. 3.7). The converse is easy. The case $u_r < u_\ell$ is parallel. $\qquad\square$

If $f$ is convex, this condition is equivalent to saying that $f'(u_r) < s < f'(u_\ell)$ for $u_r < u_\ell$. This is a classical entropy condition for convex $f$. In the case of concave $f$

**Fig. 3.7** Profile of $f$ on $(u_\ell, u_r)$

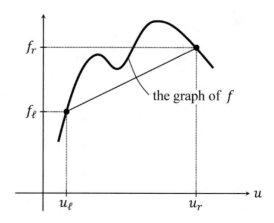

the graph of $f$

like the traffic flow problem (3.5), if $u_r < u_\ell$, then $u_S$ does NOT fulfill the traveling wave entropy condition.

In the next section, we discuss the Kružkov entropy solution, which combines such an entropy condition and the notion of a weak solution, so that one can check the entropy condition for a general function whose jump (shock) curves are not regular.

In Proposition 3.5, we discuss an equivalent condition when $u_S$ satisfies the traveling wave entropy solution. One can write this equivalent condition in a synthetic way as

$$s|k - u_\ell| < \text{sgn}(k - u_\ell)\,(f(k) - f(u_\ell))$$

for all $k$ between $u_\ell$ and $u_r$. Here, sgn denotes the *sign function* defined by

$$\text{sgn}\,x = \begin{cases} 1, & x > 0, \\ 0, & x = 0, \\ -1, & x < 0. \end{cases}$$

(Of course, one may replace $u_\ell$ with $u_r$ in the above inequality.) One may write this another way to express the condition similarly. Let $a = a(x, t)$ be a function defined in $D = J \times (t_0, t_1)$, where $J$ is an open interval in $\mathbf{R}$. Let $[\![a]\!]$ denote the difference between the limit from $D_r$ and $D_\ell$, i.e.,

$$[\![a]\!](x, t) := a_r(x, t) - a_\ell(x, t), \quad (x, t) \in \Gamma = \overline{D_r} \cap \overline{D_\ell},$$

$$a_r(x, t) = \lim \left\{ a(y, s) \mid (y, s) \to (x, t),\ (y, s) \in D_r \right\},$$

$$a_\ell(x, t) = \lim \left\{ a(y, s) \mid (y, s) \to (x, t),\ (y, s) \in D_\ell \right\}.$$

---

**Proposition 3.6**
*Let $f \in C^1(\mathbf{R})$. Consider the Riemann problem. The function $u_S$ in (3.10) satisfies the traveling wave entropy condition if and only if*

$$s[\![|u - k|]\!] \geq [\![\text{sgn}(u - k)\,(f(u) - f(k))]\!]\ \text{for all } k \in (u_\ell, u_r)\ \text{if } u_\ell < u_r$$

$$(\text{resp. } k \in (u_r, u_\ell)\ \text{if } u_r < u_\ell)\ \text{and}$$

$$s[\![|u - k|]\!] = [\![\text{sgn}(u - k)\,(f(u) - f(k))]\!]\ \text{for all } k \notin (u_\ell, u_r)\ \text{if } u_\ell < u_r$$

$$(\text{resp. } k \notin (u_r, u_\ell)\ \text{if } u_r < u_\ell)$$

*for $u = u_S$, where $x(t) = st$ with $s = (f_\ell - f_r)/(u_\ell - u_r)$, $f_\ell = f(u_\ell)$, $f_r = f(u_r)$.*

*Proof.* We only give a proof when $u_\ell < u_r$ since the proof for $u_\ell > u_r$ is symmetric. "If" part. Choosing $k$ between $u_\ell$ and $u_r$, we obtain

$$s\left(-(u_\ell - k) - (u_r - k)\right) < -\left(f_r - f(k)\right) - \left(f_\ell - f(k)\right)$$

or

$$\overline{f} + s(k - \overline{u}) < f(k).$$

Here, $\overline{f} = (f_r + f_\ell)/2, \overline{u} = (u_r + u_\ell)/2$. This implies that the graph of $f(u)$ must lie above the straight line segment between $(u_\ell, f_\ell)$ and $(u_r, f_r)$. Proposition 3.5 now implies that $u_S$ satisfies the traveling wave entropy condition.

"Only if" part. Since the Rankine–Hugoniot condition holds,

$$s[\![|u - k|]\!] = [\![\operatorname{sgn}(u - k)\left(f(u) - f(k)\right)]\!]$$

for any constant $k$ not between $u_\ell$ and $u_r$. For constants $k$ between $u_\ell$ and $u_r$, if the traveling wave entropy condition holds, then, by Proposition 3.5, we have

$$f(k) > s(k - u_\ell) + f(u_\ell) \quad \text{and}$$
$$f(k) > s(k - u_\ell) + f(u_r),$$

so that

$$f(k) - sk > \overline{f} - s\overline{u}.$$

Then we obtain

$$s[\![|u - k|]\!] > [\![\operatorname{sgn}(u - k)\left(f(u) - f(k)\right)]\!].$$

$\square$

▶ **Remark 3.7** This proposition says that for the Riemann problem, the solution satisfying the traveling entropy condition is exactly the Kružkov entropy solution defined later.

## 3.2 Uniqueness of Entropy Solutions

We first derive two equivalent definitions of an entropy solution. One is based on what we call an entropy pair, and the other is its modification due to Kružkov. The first condition is easily motivated by a vanishing viscosity approximation. We derive this condition by a formal argument. Then we introduce Kružkov's entropy condition and discuss the equivalence of both definitions. We conclude this section by proving the uniqueness of an entropy solution. The key idea is a kind of doubling variable argument.

## 3.2.1   Vanishing Viscosity Approximations and Entropy Pairs

We consider the initial value problem of a scalar conservation law of the form

$$u_t + f(u)_x = 0 \qquad\qquad \text{in} \quad Q = \mathbf{T} \times (0, T), \qquad\qquad (3.16)$$

$$u|_{t=0} = u_0 \qquad\qquad \text{on} \quad \mathbf{T}. \qquad\qquad (3.17)$$

Here $u = u(x, t)$ is a real-valued function on $Q$. In other words, to simplify the presentation, $u$ is periodic in $x$. The flux function $f$ is always assumed to be at least a locally Lipschitz (real-valued) function.

To obtain a solution, we consider a parabolic approximation

$$u_t^\varepsilon + f(u^\varepsilon)_x = \varepsilon u_{xx}^\varepsilon, \qquad\qquad (3.18)$$

$$u^\varepsilon \big|_{t=0} = u_0 \qquad\qquad (3.19)$$

for $\varepsilon > 0$. We expect that a reasonable solution will be obtained as a limit of the solution of (3.18), (3.19) as $\varepsilon \to 0$. Since $\varepsilon$ looks like a viscosity coefficient in fluid dynamics, this approximation is often called a vanishing viscosity approximation.

It is well known that (3.18) and (3.19) admit a global solution $u^\varepsilon$ that is smooth for $t > 0$ for any given $u_0 \in L^\infty(\mathbf{T})$ provided that $f$ is smooth. For a moment we assume that $f$ is smooth, so that $u^\varepsilon$ is smooth for $t > 0$ (the initial condition should be understood in a weak sense); see, for example, standard monographs on parabolic equations [70, 72]. We take a real-valued smooth function $\eta$ defined on $\mathbf{R}$, and consider a composite function $\eta(u^\varepsilon) = \eta \circ u^\varepsilon$. Since $u^\varepsilon$ satisfies (3.18), $\eta(u^\varepsilon)$ must solve

$$\eta(u^\varepsilon)_t + \eta'(u^\varepsilon) f'(u^\varepsilon) u_x^\varepsilon = \varepsilon \eta'(u^\varepsilon) u_{xx}^\varepsilon. \qquad\qquad (3.20)$$

Since $\eta(u^\varepsilon)_{xx} = \eta'(u^\varepsilon) u_{xx}^\varepsilon + \eta''(u^\varepsilon)(u_x^\varepsilon)^2$, we observe that $\eta(u^\varepsilon)_{xx} \geq \eta'(u^\varepsilon) u_{xx}^\varepsilon$ provided that $\eta$ is *convex*.

Assume that $\eta$ is now convex, and take a function $q$ such that $q' = \eta' f'$. Then (3.20) yields

$$\eta(u^\varepsilon)_t + q(u^\varepsilon)_x = \varepsilon \eta'(u^\varepsilon) u_{xx}^\varepsilon \leq \varepsilon \eta(u^\varepsilon)_{xx}. \qquad\qquad (3.21)$$

We multiply (3.21) by a nonnegative function $\varphi \in C_c^\infty(Q_0)$ on $Q_0 = \mathbf{T} \times [0, T)$ and integrate by parts to get

$$\int_Q \{\varphi_t \eta(u^\varepsilon) + \varphi_x q(u^\varepsilon)\} \, dx dt + \int_{\mathbf{T}} \varphi|_{t=0} \, \eta(u_0) dx \geq -\varepsilon \int_Q \varphi_{xx} \eta(u^\varepsilon) dx dt.$$

Here, we present a formal argument. The following argument can be justified if, for example, $\sup_Q |u^\varepsilon|$ is bounded in $\varepsilon$ and if $u^\varepsilon$ tends to $u$ almost everywhere (a.e.) in $Q$ as $\varepsilon \downarrow 0$. Sending $\varepsilon$ to zero we get

$$\int_Q \{\varphi_t \eta(u) + \varphi_x q(u)\} \, dx \, dt + \int_{\mathbf{T}} \varphi|_{t=0} \, \eta(u_0) dx \geq 0 \qquad (3.22)$$

for any $\varphi \in C_c^\infty(Q_0)$ with $\varphi \geq 0$. In $Q$ this condition implies

$$\eta(u)_t + q(u)_x \leq 0 \qquad (3.23)$$

in a distribution sense, which means $-\eta(u)_t - q(u)_x$ is a nonnegative Radon measure in $Q$.

This argument can be extended when $\eta$ is merely convex by an approximation for (3.22). (Incidentally, the inequality for taking $\eta(n)_{xx} \geq \eta'(u)u_{xx}$ for $\eta(u) = |u|$ is known as the Kato inequality $\Delta|w| \geq (\text{sgn } w)\Delta w$ in a distribution sense. This inequality is also obtained by approximately $|u|$, by, for example, $\sqrt{|u|^2 + \delta}, \delta > 0$. See Exercise 3.8.)

Inequality (3.23) is trivially fulfilled if $u$ solves (3.16) and $u$ is *smooth*. However, it will turn out that this inequality distinguishes admissible jumps and nonadmissible jumps when $u$ is discontinuous. We thus reach the following definition.

---

**Definition 3.8**

Let $f$ be a locally Lipschitz function on $\mathbf{R}$.

**(1)** A pair of functions $(\eta, q)$ is an *entropy pair* for (3.16) if $\eta$ is convex and $q$ is a primitive (antiderivative) of $\eta' f'$, i.e., $q' = \eta' f'$.

**(2)** Let $u \in L^\infty(Q)$ be a weak solution of (3.16), (3.17) with initial datum $u_0 \in L^\infty(\mathbf{T})$. Let $(\eta, q)$ be an entropy pair for (3.16). We say that $u$ is an *entropy solution* of (3.16), (3.17) if $u$ satisfies (3.22) for all $\varphi \in C_c^\infty(Q_0)$, with $\varphi \geq 0$, where $Q_0 = \mathbf{T} \times [0, T)$.

---

### 3.2.2 Equivalent Definition of Entropy Solution

For a convex function $\eta$ on $\mathbf{R}$, $(\eta, q)$ is an entropy pair (for (3.16)) if we set

$$q(w) = \int_k^w \eta'(\tau) f'(\tau) d\tau.$$

The function $q$ is uniquely determined by $\eta$ up to an additive constant. If we take $\eta(w) = |w - k|$ for $k \in \mathbf{R}$, then we have

$$q(w) = \text{sgn}(w - k) \left( f(w) - f(k) \right).$$

It is clear that if $u$ is an entropy solution, then it must satisfy

$$\int_Q \{\varphi_t |u - k| + \varphi_x \, (\mathrm{sgn}(u - k) \, (f(u) - f(k)))\} \, dxdt + \int_{\mathbf{T}} \varphi|_{t=0} \, |u_0 - k| dx \geq 0$$
(3.24)

for all $k \in \mathbf{R}$ and all $\varphi \in C_c^\infty(Q_0)$, with $\varphi \geq 0$. This condition is often called the *Kružkov entropy condition*. It is equivalent to the definition of an entropy solution.

**Proposition 3.9**
*Let $f$ be a locally Lipschitz function. Let $u \in L^\infty(Q)$ be a weak solution of (3.16), (3.17) with initial datum $u_0 \in L^\infty(\mathbf{T})$. Then $u$ is an entropy solution if and only if $u$ satisfies the Kružkov entropy condition, i.e., (3.24) for all $k \in \mathbf{R}$ and for all $\varphi \in C_c^\infty(Q_0)$, with $\varphi \geq 0$.*

*Proof.* Since the "only if" part is trivial, we shall prove the "if" part. For $\eta$ we set a linear functional

$$\Lambda(\eta) = \int_Q \{\varphi_t \eta(u) + \varphi_x q(u)\} \, dxdt + \int_{\mathbf{T}} \varphi|_{t=0} \, \eta(u_0) dx$$

for a fixed $\varphi \in C_c^\infty(Q_0)$ and $u_0$. This quantity $\Lambda(\eta)$ is determined by $\eta$ and is independent of the choice of $q$ provided that $(\eta, q)$ is an entropy pair. The Kružkov entropy condition (3.24) implies

$$\Lambda(\eta_i) \geq 0$$

for all $\eta_i(w) = \alpha_i |w - k_i|, \, k_i \in \mathbf{R}, \, \alpha_i \geq 0, \, i = 1, \cdots, m$. Thus,

$$\Lambda \left( \sum_{i=1}^m \eta_i \right) = \sum_{i=1}^m \Lambda(\eta_i) \geq 0$$

since $(\sum_{i=1}^m \eta_i, \sum_{i=1}^m q_i)$ is an entropy pair if $(\eta_i, q_i)$ is an entropy pair. Since $u$ is a weak solution, we see that $\Lambda(\eta) = 0$ if $\eta(w) = \alpha w + \beta, \, \alpha, \beta \in \mathbf{R}$. Thus, the convex piecewise linear function $\eta$ of the form

$$\eta(w) = \alpha w + \beta + \sum_{i=1}^m \eta_i(w)$$
(3.25)

satisfies $\Lambda(\eta) \geq 0$.

As stated at the end of this subsection (Lemma 3.10), we notice that any piecewise linear convex function is of the form (3.25) provided that there is only a finite number of nondifferentiable points. We thus conclude that $\Lambda(\eta) \geq 0$ for any piecewise linear convex function $\eta$. Since a convex function $\eta$ is approximable (Exercise 3.5) by a piecewise linear convex function $\{\zeta_j\}_{j=1}^{\infty}$ (having finitely many nondifferentiable points) locally uniformly in $\mathbf{R}$, we conclude that

$$\Lambda(\eta) = \lim_{j \to \infty} \Lambda(\zeta_j) \geq 0$$

since $u$ is bounded. $\qquad\square$

---

**Lemma 3.10**

*Let $\eta$ be a piecewise linear convex function in $\mathbf{R}$ with $m$ nondifferentiable points. Then there are $\alpha_i \geq 0$, $\alpha_i$, $\beta_i$, $k_i \in \mathbf{R}$ for $1 \leq i \leq m$ such that*

$$\eta(w) = \alpha w + \beta + \sum_{i=1}^{m} \eta_i(w), \quad \eta_i(w) = \alpha_i |w - k_i| + \beta_i.$$

---

*Proof.* This can be easily proved by induction of numbers $m$ of nondifferentiable points of a piecewise linear convex function $\xi_m$. If $m = 0$, it is trivial. Let $\{k_i\}_{i=1}^{m}$ be the set of all nondifferentiable points of $\xi_m$. We may assume that $k_1 < k_2 < \cdots < k_m$. Assume that $m \geq 1$. Taking $\alpha$, $\beta$, and $\alpha_1$ in a suitable way, we see that

$$\xi_m(w) = \alpha w + \beta + \eta_1(w) \quad \text{for} \quad -\infty < w < k_2,$$

where $k_2$ is the second smallest nondifferentiable point of $\xi_m$; $k_2 = \infty$ if there is no such point (Fig. 3.8).

**Fig. 3.8** Profile of graphs

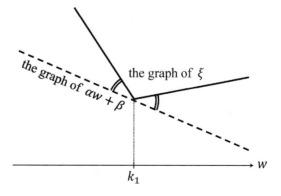

We set

$$\xi(w) = \alpha w + \beta + \eta_1(w) \quad \text{for} \quad w \in \mathbf{R}.$$

Since $\xi_m$ is convex and $\xi$ is linear for $s > k_1$, $\xi_m - \xi$ is still convex and nonnegative and $\xi_m - \xi = 0$ on $(-\infty, k_2)$. Moreover, the number of nondifferentiable points of $\xi_m - \xi$ is $m - 1$, so by induction we conclude that $\xi_m$ is of the form of (3.25).  $\square$

### 3.2.3  Uniqueness

We are now in a position to state our main uniqueness result as an application of the $L^1$-contraction property.

---

**Theorem 3.11**

*Assume that $f$ is locally Lipschitz. Let $u$ and $v (\in L^\infty(Q))$ be an entropy solution of (3.16), (3.17) with initial datum $u_0$ and $v_0$, respectively. Assume that $u(\cdot, t) \to u_0$ and $v(\cdot, t) \to v_0$ as $t \to 0$ in the sense of $L^1$-convergence. Then*

$$\|u - v\|_{L^1(\mathbf{T})}(t) \leq \|u_0 - v_0\|_{L^1(\mathbf{T})}. \tag{3.26}$$

*In particular, a bounded entropy solution of (3.16), (3.17) is unique. (The assumption of $L^1$-continuity as $t \downarrow 0$ is unnecessary but we assume it to simplify the proof.)*

---

*Proof.* We double the variables of a test function $\varphi$. Let $\phi = \phi(x, t, y, s)$ be a nonnegative function such that $\phi \in C_c^\infty(Q_0 \times Q_0)$. Since $u$ is an entropy solution of (3.16), (3.17), the Kružkov entropy condition implies

$$\int_Q \{|u - k|\phi_t + q(u, k)\phi_x\} \, dx dt + \int_{\mathbf{T}} \phi(x, 0, y, s)|u_0 - k| dx \geq 0$$

when $q(u, k) = \text{sgn}(u - k)(f(u) - f(k))$. Plugging in $k = v(y, s)$ and integrating in $(y, s)$, we get

$$\int_Q \int_Q \{|u(x, t) - v(y, s)| \, \phi_t + q \, (u(x, t), v(y, s)) \, \phi_x\} \, dx dt dy ds$$

$$+ \int_Q \int_{\mathbf{T}} \phi(x, 0, y, s) \, |u_0(x) - v(y, s)| \, dx dy ds \geq 0.$$

The same inequality holds for $v$; in other words, we have

$$\int_Q \int_Q \left\{ |u(x,t) - v(y,s)| \, \phi_s + q \, (u(x,t), v(y,s)) \, \phi_y \right\} dx dt dy ds$$

$$+ \int_T \int_Q \phi(x, 0, y, 0) \, |v_0(y) - u(x,t)| \, dx dt dy \geq 0.$$

Adding these two inequalities yields

$$\int_Q \int_Q \left\{ |u(x,t) - v(y,s)| \, (\phi_t + \phi_s) + q(u,v)(\phi_x + \phi_y) \right\} dx dt dy ds$$

$$+ \int_Q \int_T |u_0(x) - v(y,s)| \, \phi(x, 0, y, s) \, dx dy ds$$

$$+ \int_T \int_Q |u(x,t) - v_0(y)| \, \phi(x, t, y, 0) \, dx dt dy \geq 0. \qquad (3.27)$$

Our strategy is as follows. We set

$$J_1 := \int_Q \int_Q \left\{ |u(x,t) - v(y,s)| \, (\phi_t + \phi_s) + q(u,v)(\phi_x + \phi_y) \right\} dx dt dy ds$$

$$J_2 := \int_Q \int_T |u_0(x) - v(y,s)| \, \phi(x, 0, y, s) \, dx dy ds$$

$$J_3 := \int_T \int_Q |u(x,t) - v_0(y)| \, \phi(x, t, y, 0) \, dx dt dy.$$

For a given $t_0 \in (0, T)$, we would like to take a suitable $\phi$ so that $J_i$ equals $I_i$ $(i = 1, 2, 3)$, with

$$I_1 := - \int_T |u(x,t_0) - v(x,t_0)| \, dx$$

$$I_2 := \frac{1}{2} \int_T |u_0(x) - v_0(x)| \, dx, \quad I_3 := I_2.$$

Since (3.27) says $J_1 + J_2 + J_3 \geq 0$, we have

$$I_1 + I_2 + I_3 \geq 0, \quad \text{i.e.,}$$

$$- \int_T |u(x,t_0) - v(x,t_0)| \, dx + \frac{1}{2} \int_T |u_0(x) - v_0(x)| \, dx$$

$$+ \frac{1}{2} \int_T |u_0(x) - v_0(x)| \, dx \geq 0. \qquad (3.28)$$

This is simply the desired contraction property (3.26). Unfortunately, there is no good function $\phi$. We need a sequence $\phi = \phi_{\varepsilon,\varepsilon',\varepsilon''}$ depending on three parameters $\varepsilon, \varepsilon', \varepsilon'' > 0$. Let $\rho_\varepsilon$ be a Friedrichs' mollifier $\rho_\varepsilon$ defined in §2.2 (see also Lemma 3.12 in what follows). We further assume symmetry, i.e., $\rho_\varepsilon(-\sigma) = \rho_\varepsilon(\sigma)$ for all $\sigma \in \mathbf{R}$. We set

$$\phi(x, t, y, s) = \rho_\varepsilon(x - y)\rho_{\varepsilon'}(t - s)\chi_{\varepsilon''}\left(\frac{t + s}{2} - t_0\right),$$

with $\chi_{\varepsilon''}(\tau) = \int_\tau^\infty \rho_{\varepsilon''}(\sigma)d\sigma$. (We shall give a heuristic explanation as to why this choice is good after the proof.) It suffices to prove

$$\lim_{\varepsilon'' \downarrow 0}\left(\lim_{\varepsilon' \downarrow 0}\left(\lim_{\varepsilon \downarrow 0} J_i(\varepsilon, \varepsilon', \varepsilon'')\right)\right) = I_i \quad \text{for} \quad i = 1, 2, 3,$$

to get (3.28) since $J_1 + J_2 + J_3 \geq 0$.

Since $\phi_x + \phi_y = 0$ and

$$\phi_t + \phi_s = \rho_\varepsilon(x - y)\rho_{\varepsilon'}(t - s)\chi'_{\varepsilon''}\left(\frac{t + s}{2} - t_0\right),$$

we observe that

$$J_1 = J_1(\varepsilon, \varepsilon', \varepsilon'')$$

$$= -\int_Q\int_Q |u(x, t) - v(y, s)|\, \rho_\varepsilon(x - y)\rho_{\varepsilon'}(t - s)\rho_{\varepsilon''}\left(\frac{t + s}{2} - t_0\right)dxdtdyds.$$

We apply the approximation lemma (Lemma 3.12 below) to conclude that

$$\lim_{\varepsilon'' \downarrow 0}\left(\lim_{\varepsilon' \downarrow 0}\left(\lim_{\varepsilon \downarrow 0} J_1\right)\right) = I_1.$$

Similarly,

$$\lim_{\varepsilon \downarrow 0} J_2 = \int_0^T\int_{\mathbf{T}} |u_0(x) - v(x, s)|\, \rho_{\varepsilon'}(s)\chi_{\varepsilon''}\left(\frac{s}{2} - t_0\right)dxds.$$

For a given $t_0 \in (0, T)$, we take $\varepsilon'' > 0$ small, say, $\varepsilon'' < \varepsilon_0''$, for some $\varepsilon_0'' > 0$, so that $\chi_{\varepsilon''}\left(\frac{s}{2} - t_0\right) = 1$ for all $s \in [0, t_0/2]$, $\varepsilon'' < \varepsilon_0''$. We take $\varepsilon' > 0$ sufficiently small so that supp $\rho_{\varepsilon'} \subset [0, t_0/2]$ to get

$$\lim_{\varepsilon \downarrow 0} J_2 = \int_0^T \left\{\int_{\mathbf{T}} |u_0(x) - v(x, s)|\, dx\right\} \rho_{\varepsilon'}(s)\, ds.$$

Since we have assumed that $v(\cdot, t) \to v_0$ in $L^1(\mathbf{T})$,

$$h(s) = \int_{\mathbf{T}} |u_0(x) - v(x, s)| \, dx = \|u_0 - v(\cdot, s)\|_{L^1(\mathbf{T})}$$

is continuous at $s = 0$. We now apply Lemma 3.12 (ii) to conclude that

$$\lim_{\varepsilon' \downarrow 0} \left( \lim_{\varepsilon \downarrow 0} J_2 \right) = \frac{1}{2} h(0) = I_1$$

for $\varepsilon'' < \varepsilon_0''$. The proof for $J_3$ is the same. We now conclude that

$$\lim_{\varepsilon'' \downarrow 0} \left( \lim_{\varepsilon' \downarrow 0} \left( \lim_{\varepsilon \downarrow 0} J_i \right) \right) = I_i$$

so that $J_1 + J_2 + J_3 \geq 0$ implies (3.28). The proof is now complete. □

Let us say a few words about why we choose $\phi$ as earlier. It is convenient to use what is called a delta function $\delta$. It is defined as a distributional derivative of a Heaviside function $1_{>0}$, i.e.,

$$\delta = D_x 1_{>0},$$

where $1_{>0}(x) = 1$ if $x > 0$ and $1_{>0}(x) = 0$ if $x \leq 0$. In other words,

$$\delta(\varphi) = -\int_{\mathbf{R}} \frac{d\varphi}{dx} 1_{>0} \, dx \quad \text{for} \quad \varphi \in C_c^\infty(\mathbf{R}).$$

By definition, $\delta(\varphi) = -\int_0^\infty \frac{d\varphi}{dx} dx = \varphi(0)$. We often write $\delta(\varphi)$ by $\int_{\mathbf{R}} \delta(x) \varphi(x) dx$, though $\delta$ cannot be identified with any integrable function. We would like to take

$$\phi(x, t, y, s) = \delta(x - y) \delta(t - s) 1_{>0} \left( t_0 - \frac{t + s}{2} \right).$$

Since $\phi_x + \phi_y = 0$, $\phi_t + \phi_s = -\delta(x - y)\delta(t - s)\delta(t - t_0)$, we see that

$$J_1 = -\int_{\mathbf{R}} |u(x, t_0) - v(x, t_0)| \, dx = I_1.$$

Since $u$ and $v$ are not necessarily continuous, we must approximate $\delta$ by mollifiers. For $J_2 + J_3$, we have

$$J_1 + J_2 = \int_0^{t_0} \left\{ \int_{\mathbf{T}} |u_0(x) - v(x, s)| \, dx \right\} \delta(-s) ds$$

$$+ \int_0^{t_0} \left\{ \int_{\mathbf{T}} |v_0(x) - u(x, s)| \, dx \right\} \delta(t) \, dt$$

$$= \int_{-t_0}^{t_0} k(t) \delta(t) \, dt = k(0),$$

with

$$k(t) = \begin{cases} \int_{\mathbf{T}} |v_0(x) - u(x, t)| \, dx & \text{for } t > 0, \\ \int_T |u_0(x) - v(x, -t)| \, dx & \text{for } t \leq 0. \end{cases}$$

Since $k$ is continuous at $t = 0$ and $k(0) = \|u_0 - v_0\|_{L^1(\mathbf{T})}$, we observe that

$$I_1 + \|u_0 - v_0\|_{L^1(\mathbf{T})} \geq 0.$$

In our proof, we discuss $J_1$ and $J_2$ separately, so we use symmetry to simplify the argument.

**Lemma 3.12**

*Let $\rho_\varepsilon$ be a Friedrichs' mollifier on $\mathbf{R}$ defined in Sect. 2.2. In other words, $\rho_\varepsilon(\sigma) = \varepsilon^{-1}\rho(\sigma/\varepsilon)$, where $\rho \in C_c^\infty(\mathbf{R})$ satisfies $\rho \geq 0$ and $\int_{\mathbf{R}} \rho dx = 1$.*

**(i)** *Let $h \in L^\infty(\mathbf{T}^2)$ and $h(x, x - z) \to h(x, x)$ as $|z| \to 0$ for a.e. $x$. Then*

$$\lim_{\varepsilon \downarrow 0} \int_{\mathbf{T}} \int_{\mathbf{T}} h(x, y) \rho_\varepsilon(x - y) \, dx dy = \int_{\mathbf{T}} h(x, x) \, dx.$$

*Let $h \in L^\infty(\mathbf{R})$ be compactly supported. Assume that $h(x, x - z) \to h(x, x)$ as $|z| \to 0$ for a.e. $x$. Then*

$$\lim_{\varepsilon \downarrow 0} \int_{\mathbf{R}} \int_{\mathbf{R}} h(x, y) \rho_\varepsilon(x - y) \, dx dy = \int_{\mathbf{R}} h(x, x) \, dx.$$

**(ii)** *Assume further that $\rho(-\sigma) = \rho(\sigma)$ for $\sigma \in \mathbf{R}$. For $h \in L^\infty(0, T)$,*

$$\lim_{\varepsilon \downarrow 0} \int_0^T h(s) \rho_\varepsilon(s) ds = \frac{1}{2} h(0)$$

*provided that $h$ is continuous at $s = 0$.*

*Proof of Lemma 3.12*

(i) We may assume supp $\rho \subset (-1, 1)$ by replacing $\sigma$ with $\sigma/\varepsilon'$ for small $\varepsilon' > 0$. We recall $\mathbf{T} = \mathbf{R}/\omega_1\mathbf{Z}$. We take $\varepsilon < \omega_1/4$, so that supp $\rho \subset \left(-\frac{\omega_1}{4\varepsilon}, \frac{\omega_1}{4\varepsilon}\right)$. By this choice, the support of $\rho_\varepsilon (x - y)$ as a function of $x$, $y$ is contained in a periodic cell $C = [-\omega_1/2, \omega_1/2) \times [-\omega_1/2, \omega_1/2)$ of $\mathbf{T}^2 = (\mathbf{R}/\omega_1\mathbf{Z})^2$. In particular,

$$\int_{-\omega_1/2}^{\omega_1/2} \rho_\varepsilon(x - y)\, dy = 1 \quad \text{for} \quad x \in (-\omega_1/2, \omega_1/2).$$

We proceed with

$$I(\varepsilon) := \int_{\mathbf{T}}\int_{\mathbf{T}} h(x, y)\rho_\varepsilon(x - y)\, dx dy - \int_{\mathbf{T}} h(x, x)\, dx$$

$$= \iint_C (h(x, y) - h(x, x))\, \rho_\varepsilon(x - y)\, dx dy.$$

Changing the variables of integration from $(x, y)$ to $(x, z)$ with $z = (x - y)/\varepsilon$, we obtain, by Fubini's theorem, that

$$|I(\varepsilon)| \le \iint_C |h(x, y) - h(x, x)|\, \rho_\varepsilon(x - y)\, dx dy$$

$$= \int_{|x|\le\omega_1/2} \left\{ \int_{|x-\varepsilon z|\le\omega_1/2} |h(x, x - \varepsilon z) - h(x, x)|\, \rho(z)\, dx \right\} dz$$

$$\le \int_{|x|\le\omega_1/2} \left\{ \int_{|z|\le 1} |h(x, x - \varepsilon z) - h(x, x)|\, \rho(z)\, dx \right\} dz.$$

Since the integrand is bounded by $2\|h\|_\infty$ (independent of $\varepsilon > 0$) and $h(x, x - \varepsilon z) \to h(x, x)$ for a.e. $x, z \in \mathbf{R}$ as $\varepsilon \downarrow 0$, we conclude that $I(\varepsilon) \to 0$ as $\varepsilon \to 0$ by the dominated convergence theorem. We thus obtained the first statement. The proof for the second statement is parallel.

(ii) By a change of the variable of integration, we see that

$$\int_0^T h(s)\rho_\varepsilon(s)\, ds = \int_{0\le z\le 1} h(\varepsilon z)\rho(z)\, dz$$

for sufficiently small $\varepsilon > 0$. Since $h(z) \to h(0)$ as $z \to 0$, we now obtain

$$\lim_{\varepsilon\downarrow 0} \int_0^T h(s)\rho_\varepsilon(s)\, ds = \lim_{\varepsilon\downarrow 0} \int_{0\le z\le 1} h(\varepsilon z)\rho(z)\, dz = h(0) \int_{0\le z\le 1} \rho(z)\, dz$$

by the bounded convergence theorem. The result follows if we note that
$\int_0^\infty \rho(z)dz = 1/2$ by symmetry.                                        □

► **Remark 3.13**

(i) The definition of the entropy solution extends to a bounded function in **R** not
necessarily periodic. Although the uniqueness result still holds provided that $u_0$
is in $L^1(\mathbf{R}) \cap L^\infty(\mathbf{R})$, the proof is more involved. For example, we must take $\phi$
to have compact support in the space direction; See, for example, [53] and [68].

(ii) All results here can be extended to multidimensional space. The conservation
law for a real-valued function $u = u(x, t)$ is of the form

$$u_t + \operatorname{div} f(u) = 0 \quad \text{in} \quad \mathbf{T}^N \times (0, T) = \tilde{Q},$$

with $f(u) = (f^1(u), \ldots, f^N(u))$. A pair of real-valued functions $(\eta, q)$
defined on **R** is said to be an *entropy pair* for this equation if it satisfies
$q' = \eta' f'$ $(i = 1, \ldots, N)$, $q = (q^1, \ldots, q^N)$ and $\eta$ is a convex function.
A function $u \in L^\infty(\tilde{Q})$ is said to be an entropy solution with initial datum
$u_0 \in L^\infty(\mathbf{T}^N)$ if

$$\int_0^T \!\!\int_{\tilde{Q}} \left( \varphi_t \eta(u) + \sum_{i=1}^N q^i(u)\varphi_{x_i} \right) \mathrm{d}x\mathrm{d}t + \int_{\mathbf{T}^N} \varphi|_{t=0}\, \eta(u_0)\, \mathrm{d}x \geq 0$$

holds for all $\varphi \in C_c^\infty(\tilde{Q}_0)$, with $\varphi \geq 0$, and all entropy pairs. Here $\tilde{Q}_0 = \mathbf{T}^N \times [0, T)$ and $\varphi_{x_i} = \frac{\partial \varphi}{\partial x_i}$. The Kružkov entropy condition is of the form

$$\int_0^T \!\!\int_{\tilde{Q}} \left( \varphi_t |u - k| + \sum_{i=1}^N \operatorname{sgn}(u - k) \left( f^i(u) - f^i(k) \right) \varphi_{x_i} \right) \mathrm{d}x\mathrm{d}t$$

$$+ \int_{\mathbf{T}^N} \varphi|_{t=0} |u - k|\, \mathrm{d}x \geq 0$$

for all $\varphi \in C_c^\infty(\tilde{Q}_0)$ with $\varphi \geq 0$ and $k \in \mathbf{R}$; see, for example, [68].

## 3.3    Notes and Comments

Most of the contents in this chapter are taken from Holden and Risebro's book
[53], where **T** is replaced by **R**. The theory of conservation laws has a long
history. A weak formulation for the Burgers equation traces back to Hopf [54],
where a parabolic approximation was studied. The literature on the topic has grown
considerably since then. The reader is referred to the book [53].

There are several ways to construct an entropy solution, for example, [53]. Of course, parabolic approximation is one way. Other methods are based on the finite difference method. A front tracking method was studied extensively by Holden and Risebro [53]; it approximates $f$ by a piecewise function; this seems to be very effective even for systems of conservation laws. A completely different approach, called a kinetic construction (not contained in [53]), traces back to Brenier [14], as well as the second author and others [49], [50]. The idea involves introducing an extra variable, which may be interpreted as a microscopic variable. All the aforementioned methods work for scalar conservation laws in multidimensional spaces. Note that there is a very accessible introduction to conservation laws in the book [36, Chapter 11]. In [36], systems of conservation laws are discussed.

If one considers systems of conservation laws, the uniqueness of entropy solutions is difficult because there are interactions of waves. Neverthless, there are now several uniqueness results that go back to Bressan's seminal works [16], [17], where the main assumption is that the spatial total variation of a solution is small. The reader is referred to [17] or [53] for this topic.

## 3.4 Exercises

3.1 (Hopf–Cole transformation) Let $u$ be a solution of the (viscous) Burgers equation $u_t + (u^2/2)_x = u_{xx}$. Let $w(x, t)$ be defined as

$$w(x, t) = \int_0^x u(y, t) \mathrm{d}y + \int_0^t \left( u_x(0, \tau) - u(0, t)^2/2 \right) \mathrm{d}\tau.$$

Show that $w$ satisfies

$$w_t + (w_x)^2/2 = w_{xx}$$

in $\mathbf{R} \times (0, \infty)$. Show that $v = \exp(-w/2)$ solves the heat equation $v_t = v_{xx}$.

3.2 Let $u$ be a solution of $u_t + (u^2/2)_x = u_{xx}$ in $\mathbf{R} \times (0, \infty)$. Set $u_\lambda(x, t) = \lambda u(\lambda x, \lambda^2 t)$ for $\lambda > 0$. Show that $u_\lambda$ solves the same equation as $u$. Set $v_\varepsilon(x, t) = v(\varepsilon x, \varepsilon t)$. Show that $v_\varepsilon$ solves $v_t + (v^2/2)_x = \varepsilon^{-1} v_{xx}$ for $\varepsilon > 0$.

3.3 Consider (3.2), with $f(u) = u^2 / \left( u^2 + (1 - u)^2 \right)$. Find the entropy solution to the Riemann problem with initial datum (3.9), where $u_\ell = 0$, $u_r = 1$. In this case, the equation is called the Buckley–Leverett equation. It is a simple model of two-phase fluid flow in a porous medium. The unknown $u$ represents a ratio of saturation of one of the phases. It varies from zero to one. Note that $f$ is neither convex nor concave. The expected solution has a rarefaction and shock simultaneously. Note that there is a numerical method based on the level-set approach [88] discussed in Sect. 4.5.2.

3.4 Consider an equation for $u = u(x, t)$ in $\mathbf{R} \times (0, \infty)$ of the form

$$u_t + (u^2/2)_x = -u$$

with initial datum $u_0$ in (3.9). Find the entropy solution when $u_\ell = 1$, $u_r = 0$. Consider the case where $u_\ell = 0$, $u_r = 1$. Find the entropy solution in this case.

3.5 Let $\xi$ be a real-valued convex function on $\mathbf{R}$. Prove that there exists a sequence of piecewise linear convex functions $\{\eta_j\}_{j=1}^\infty$ such that
   (i) $\eta_j$ converges to $\xi$ locally uniformly in $\mathbf{R}$ as $j \to \infty$ and
   (ii) $\eta_j$ has at most finitely many nondifferentiable points.

3.6 Let $f$ be a strictly convex $C^1$ function in the sense that $f' \in C(\mathbf{R})$ is (strictly) increasing. We set

$$u_R(x, t) = \begin{cases} u_\ell, & x < f'(u_\ell)t \\ (f')^{-1}(x/t), & f'(u_\ell)t \le x < f'(u_r)t \\ u_r, & x \ge f'(u_r)t \end{cases}$$

for $u_\ell < u_r$. Show that this is a weak solution of the Riemann problem to (3.2) with initial datum $u_0$ defined in (3.9). This solution is called a *rarefaction wave* solution. Show that $u_R$ is indeed an entropy solution by checking the Kružkov entropy condition.

3.7 Let $\xi$ be a real-valued convex function on $\mathbf{R}$. Prove that $\xi$ is Lipschitz continuous in any bounded interval $(a, b)$.

3.8 Let $u$ be a real-valued $C^2$ function on $\mathbf{R}^N$.
   (i) Let $\eta$ be a real-valued $C^2$ convex function on $\mathbf{R}$. Show that

$$\Delta \eta(u) \ge \eta'(u)\Delta u \quad \text{in} \quad \mathbf{R}^N.$$

(ii) Show that

$$\int_{\mathbf{R}^N} (\Delta\varphi)|u|\,dx \ge \int_{\mathbf{R}^N} \varphi(\operatorname{sgn} u)\Delta u\,dx$$

for any $\varphi \in C_c^\infty(\mathbf{R}^N)$ and $\varphi \ge 0$.

3.9 Let $\xi$ be a real-valued $C^2$ function on $\mathbf{R}^N$. Show that $\xi$ is convex in $\mathbf{R}^N$ if and only if its Hessian matrix $\left(\frac{\partial^2 \xi}{\partial x_i \partial x_j}(x)\right)_{1\le i,j\le N}$ is nonnegative definite for all $x \in \mathbf{R}^N$, i.e.,

$$\sum_{1\le i,j\le N} \frac{\partial^2 \xi}{\partial x_i \partial x_j}(x)z_i z_j \ge 0$$

for all $z = (z_1, \ldots, z_N) \in \mathbf{R}^N$.

3.10 Give an example of a function $f \in C\left(\mathbf{R}^2\backslash\{0\}\right)$ such that

$$a := \lim_{x\to 0}\left(\lim_{y\to 0} f(x, y)\right) \quad \text{and} \quad b := \lim_{y\to 0}\left(\lim_{x\to 0} f(x, y)\right)$$

exists but $a \ne b$.

# Hamilton–Jacobi Equations

**4**

In the last chapter, we discussed uniqueness in a special class of weak solutions called entropy solutions for scalar conservation laws, which are quasilinear first-order equations. The notion of a weak solution and an entropy solution is based on integration by parts or a variational principle.

In this chapter, we consider another class of nonlinear first-order equations whose nonlinearity is very strong and not quasilinear. Such an equation is often called the Hamilton–Jacobi equation. It is in general impossible to introduce the notion of a weak solution by integration by parts. Instead, we introduce a notion of a weak solution based on the maximum principle. Such a notion was first introduced by Crandall and Lions [29] in the early 1980s as a viscosity solution and has been extensively studied.

In this chapter, we study uniqueness problems of viscosity solutions for several types of equations. We first observe that one-dimensional evolutionary Hamilton–Jacobi equations are formally an integration of a one-dimensional scalar conservation law. We then discuss the uniqueness issue for its stationary form, the eikonal equation, as well as evolutionary Hamilton–Jacobi equations. We also discuss a scalar conservation law and its generalization from the viewpoint of viscosity solutions to handle jump discontinuities.

## 4.1 Hamilton–Jacobi Equations from Conservation Laws

In this section, we derive a fully nonlinear equation of first order called a Hamilton–Jacobi equation from a conservation law. We shall give another interpretation of the entropy condition. We also derive a kind of stationary problem, the eikonal equation.

© The Author(s), under exclusive license to Springer Nature Switzerland AG 2023    97
M.-H. Giga, Y. Giga, *A Basic Guide to Uniqueness Problems for Evolutionary Differential Equations*, Compact Textbooks in Mathematics,
https://doi.org/10.1007/978-3-031-34796-2_4

### 4.1.1   Interpretation of Entropy Solutions

We consider a conservation law for a real-valued function $u = u(x, t)$, $x \in \mathbf{R}$, $t > 0$
of the form

$$u_t + (f(u))_x = 0, \tag{4.1}$$

where $f$ is a given real-valued continuous function on $\mathbf{R}$. We integrate from $0$ to $x$
to get

$$\tilde{U}_t + f(\tilde{U}_x) = f(u(0, t)) \tag{4.2}$$

if we set $\tilde{U}(x, t) = \int_0^x u(y, t)\, dy$. We set

$$U(x, t) = \tilde{U}(x, t) - \int_0^t f(u(0, s))\, ds$$

and obtain

$$U_t + f(U_x) = 0. \tag{4.3}$$

This equation is fully nonlinear and called an (evolutionary) Hamilton–Jacobi
equation. This is simply a formal procedure since $u$ may jump at $x = 0$, so the
value $f(u(0, t))$ is not well defined.

We consider a Riemann problem for (4.1) with initial condition

$$u(x, 0) = u_0(x), \tag{4.4}$$

with

$$u_0(x) = \begin{cases} -\alpha, & x < 0, \\ \alpha, & x > 0, \end{cases}$$

where $\alpha \in \mathbf{R}$, $\alpha \neq 0$. To simplify the presentation, we set $f(u) = u^2/2$, which
corresponds to the case of the Burgers equation. As we already observed in Chap. 3,

$$u(x, t) = u_0(x)$$

is an entropy solution to (4.1) with (4.4) if $\alpha < 0$. It is not an entropy solution when
$\alpha > 0$. For $\alpha > 0$ the entropy solution is a rarefaction wave $u_R$ of the form

$$u_R(x, t) = \begin{cases} -\alpha, & x < x_\ell(t), \\ \dfrac{x}{t}, & x_\ell(t) < x < x_r(t), \\ \alpha, & x_r(t) < x, \end{cases}$$

where $x_\ell(t) = -\alpha t$ and $x_r(t) = \alpha t$. It is continuous for $t > 0$. Thus, if $\alpha > 0$, then $U = U_R$ with

$$U_R = \int_0^x u_R(y, t)\, dy$$

solves (4.3) since $u_R(0, t) = 0$. However, if $\alpha < 0$, then the term $f(u(0, t))$ should be interpreted as $f(u(+0, t))\, (= f(u(-0, t))) = f(\alpha)$. Thus,

$$\tilde{U}(x, t) = \int_0^x u(y, t)\, dy = \int_0^x u_0(y)\, dy$$

solves (4.2) with the right-hand side $f(u(0, t)) = f(\alpha)$. We thus conclude that

$$V(x, t) = \int_0^x u(y, t)dy - f(\alpha)t = v_0(x) - f(\alpha)t,$$

with

$$v_0(x) = \begin{cases} -\alpha x, & x < 0, \\ \alpha x, & x > 0, \end{cases}$$

solves (4.3) with initial datum $v_0(x)$. Although $V$ even "solves" (4.3) for $\alpha > 0$, its derivative $u$ is not an entropy solution of (4.1). We have two solutions $U_R$ and $V$ with the same initial datum $v_0$ if $\alpha > 0$. We would like to choose a solution whose spatial derivative is an entropy solution.

We recall that an entropy solution is obtained as a vanishing viscosity method. In other words, it is as a limit of the $\varepsilon$-approximated equation

$$u_t^\varepsilon + f(u^\varepsilon)_x = \varepsilon u_{xx}^\varepsilon.$$

As previously, we set

$$U^\varepsilon(x, t) = \int_0^x u^\varepsilon(y, t)\, dy - \int_0^x f\left(u^\varepsilon(0, s)\right) ds$$

and obtain

$$U_t^\varepsilon + f(U_x^\varepsilon) = \varepsilon U_{xx}^\varepsilon.$$

By the construction of an entropy solution, it is clear that our solutions $U_R$ for $\alpha > 0$ and $V$ for $\alpha < 0$ are obtained as a limit $\lim_{\varepsilon \downarrow 0} U^\varepsilon$, at least formally.

### 4.1.2   A Stationary Problem

We continue to assume that $f(u) = u^2/2$. The solution $V$ in Sect. 4.1.1 we found does not change its profile. It is a translative solution of (4.3) or a soliton-like solution. If we consider $W = V + f(\alpha)t$, then $W$ solves

$$- f(\alpha) + f(W_x) = 0. \tag{4.5}$$

This is a stationary Hamilton–Jacobi equation. For $\alpha < 0$, this solution is obtained as a limit of aforementioned vanishing viscosity approach, while for $\alpha > 0$, it is not obtained as such a limit.

Although so far we assume for simplicity that $f(u) = u^2/2$, all arguments in Sects. 4.1.1 and 4.1.2 work for a general convex function $f$ with $f(\sigma) = f(-\sigma)$ for all $\sigma \in \mathbf{R}$ and $f(0) = 0$ with modification of the explicit formula of the rarefaction wave $u_R$.

The equation $f(U_x) = g(x)$ is often called *the eikonal equation*. If $f(u) = u^2/2$, then this is of the form $|U_x| = \sqrt{2g}$. In multidimensional cases, it is of the form

$$|\nabla U| = G \quad \text{in} \quad \Omega,$$

where $G$ is a given function defined in a domain $\Omega$ in $\mathbf{R}^N$.

## 4.2     Eikonal Equation

In this section, we begin with a one-dimensional eikonal equation and then introduce a notion of viscosity solution to distinguish jumps of derivatives. We conclude this section by proving uniqueness (comparison principle) based on a kind of doubling-variables argument, unlike in Chap. 3.

### 4.2.1   Nonuniqueness of Solutions

We consider a very simple example of the eikonal equation

$$\left| \frac{du}{dx} \right| - 1 = 0 \quad \text{in} \quad (-1, 1) \tag{4.6}$$

with the Dirichlet boundary condition

$$u(\pm 1) = 0. \tag{4.7}$$

Here, $u = u(x)$ is a real-valued function defined for $x \in (-1, 1)$. It is clear that there is no $C^1$ solution. If one allows continuous functions satisfying the equation except at finitely many points, there are infinitely many solutions (even if nonnegative solutions are considered). For example,

$$u_0(x) = 1 - |x|, \qquad\qquad\qquad |x| \le 1,$$
$$u_k(x) = \frac{1}{2^k} a(2^k x), \quad k = 1, 2, \ldots, \ |x| \le 1,$$

with

$$a(y) = \max \left\{ 1 - |y - (2m + 1)| \mid m \in \mathbf{Z} \right\},$$

are such solutions (Fig. 4.1). One would like to choose a typical solution of (4.6) and (4.7). One natural solution is a distance function from the boundary $\pm 1$, which corresponds to $u_0$. See Exercise 4.8.

To conclude that a solution is unique, we must impose extra conditions like an entropy condition, which is obtained using a vanishing viscosity method. We consider for $\varepsilon > 0$

$$\left| \frac{du}{dx} \right| - 1 = \varepsilon \frac{d^2 u}{dx^2} \quad \text{in} \quad (-1, 1). \tag{4.8}$$

Then it is easy to see that (4.8) under (4.7) admits a unique $C^2$ solution $u_\varepsilon$. Indeed, it can be written as

$$u_\varepsilon(x) = \begin{cases} 1 - x + \varepsilon(e^{-1/\varepsilon} - e^{-x/\varepsilon}), & 0 \le x \le 1, \\ 1 + x + \varepsilon(e^{-1/\varepsilon} - e^{x/\varepsilon}), & -1 \le x < 0. \end{cases}$$

The uniqueness can be proved using the uniqueness of the initial value problem of ordinary differential equations in Sect. 1.1. (We do not give details here. The uniqueness can also be proved by the maximum principle for second-order ordinary differential equations; see, for example, [84].) If we take its limit as $\varepsilon \to 0$, then

**Fig. 4.1** Graphs of $u_k$

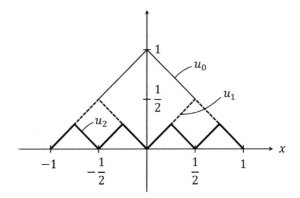

evidently $u_\varepsilon(x) \to u_0(x)$. We would like to choose $u_0$ as a "reasonable" solution of (4.6) with (4.7). It is desirable to check whether or not it is a reasonable solution without approximation. In other words, we must find a suitable notion like entropy solution to choose a reasonable solution.

### 4.2.2  Viscosity Solution

We consider a general Hamilton–Jacobi equation in a *domain* (i.e., connected open set) $\Omega$ in $\mathbf{R}^N$ of the form

$$H(x, \nabla u) = 0. \tag{4.9}$$

Here $H$ is a (real-valued) continuous function in $\overline{\Omega} \times \mathbf{R}^N$, and $\nabla u = (\partial_1 u, \ldots, \partial_N u)$ is the gradient of a scalar function $u = u(x)$, $x \in \Omega$. To motivate the definition of a viscosity solution, we consider a $C^2$ solution $u$ and consider $\varphi \in C^2(\Omega)$ such that $\max_{\overline{\Omega}}(u - \varphi) = (u - \varphi)(\hat{x})$ for some $\hat{x} \in \Omega$. We know that at the maximum point $\hat{x}$

$$\nabla(u - \varphi)(\hat{x}) = 0, \quad \nabla^2(u - \varphi)(\hat{x}) \le O$$

$$\text{or} \quad \nabla u(\hat{x}) = \nabla \varphi(\hat{x}), \quad \nabla^2 u(\hat{x}) \le \nabla^2 \varphi(\hat{x}).$$

Here, $\nabla^2 u = (\partial_{x_i} \partial_{x_j} u)$ denotes the $N \times N$ Hessian matrix of $u$ and $O$ denotes the $N \times N$ zero matrix. For two symmetric matrices $A$ and $B$, we say that $A \le B$ if the corresponding quadratic form for $B - A$ is nonnegative, i.e.,

$$\langle \eta, (B - A)\eta \rangle \ge 0$$

for all $\eta \in \mathbf{R}^N$. Let $\Delta$ denote the Laplace operator, i.e., $\Delta u = \sum_{i=1}^N \partial_i^2 u$. Assume that a solution $u$ of (4.9) is obtained as a vanishing viscosity approach, more precisely $u$ is given a limit of $u^\varepsilon$ as $\varepsilon \downarrow 0$, and $u^\varepsilon$ solves

$$H(x, \nabla u^\varepsilon) = \varepsilon \Delta u^\varepsilon.$$

Let $x_\varepsilon \in \Omega$ be a maximum point of $u^\varepsilon - \varphi$ in $\overline{\Omega}$. Assume that $x_\varepsilon \to \hat{x}$ as $\varepsilon \downarrow 0$. Then,

$$H(x_\varepsilon, \nabla \varphi(x_\varepsilon)) \le \varepsilon \Delta \varphi(x_\varepsilon)$$

since $\nabla^2 u \le \nabla^2 \varphi$ at $x = x_\varepsilon$ implies $\Delta u \le \Delta \varphi$ at $x = x_\varepsilon$; we here note that $\Delta u = \mathrm{tr}(\nabla^2 u)$ and $\Delta \varphi = \mathrm{tr}(\nabla^2 \varphi)$. Since $u$ is a limit of $u^\varepsilon$ and $x_\varepsilon \to \hat{x}$, we only obtain

$$H(\hat{x}, \nabla \varphi(\hat{x})) \le 0$$

instead of $H\left(\hat{x}, \nabla\varphi(\hat{x})\right) = 0$. Based on this observation, we arrive at the following definition of a viscosity solution.

---

**Definition 4.1**

A function $u \in C(\Omega)$ is said to be a *viscosity subsolution* of (4.9) in $\Omega$ if

$$H\left(\hat{x}, \nabla\varphi(\hat{x})\right) \leq 0$$

whenever $(\varphi, \hat{x}) \in C^1(\Omega) \times \Omega$ fulfills $\max_\Omega(u - \varphi) = (u - \varphi)(\hat{x})$. A function $u \in C(\Omega)$ is said to be a *viscosity supersolution* of (4.9) in $\Omega$ if

$$H\left(\hat{x}, \nabla\varphi(\hat{x})\right) \geq 0$$

whenever $(\varphi, \hat{x}) \in C^1(\Omega) \times \Omega$ fulfills $\min_\Omega(u - \varphi) = (u - \varphi)(\hat{x})$. If $u$ is a viscosity sub- and supersolution, then $u$ is said to be a *viscosity solution*.

---

It is easy to see that the $C^1$ function $u$ is a viscosity subsolution if and only if $u$ is a subsolution, i.e.,

$$H\left(x, \nabla u(x)\right) \leq 0 \quad \text{in} \quad \Omega.$$

We now check the example in the last subsection, where

$$H(x, \nabla u) = \left|\frac{du}{dx}\right| - 1.$$

It is easy to see that $u_k$ is a viscosity subsolution in $(-1, 1)$, but it is not a viscosity supersolution in $(-1, 1)$, except $k = 0$. Thus, among $\{u_k\}$, $u_0$ is the only viscosity solution.

Note that the notion of viscosity solution for $\left|\frac{du}{dx}\right| - 1 = 0$ and $1 - \left|\frac{du}{dx}\right| = 0$ is different. In fact, $-u_0$ is a viscosity solution of $1 - \left|\frac{du}{dx}\right| = 0$, but it is not a viscosity solution of $\left|\frac{du}{dx}\right| - 1 = 0$ (Exercise 4.1).

## 4.2.3 Uniqueness

We now consider the uniqueness problem for the eikonal equation

$$|\nabla u| - f(x) = 0 \quad \text{in} \quad \Omega. \tag{4.10}$$

Let $\partial\Omega$ denote the *boundary* of $\Omega$.

**Theorem 4.2 (Comparison principle)**
*Let $\Omega$ be a bounded domain in $\mathbf{R}^N$. Assume that $f \in C(\overline{\Omega})$ is positive in $\Omega$.
Let $u \in C(\overline{\Omega})$ and $v \in C(\overline{\Omega})$ be a viscosity sub- and supersolution of (4.10),
respectively. If $u \leq v$ on $\partial\Omega$, then $u \leq v$ in $\overline{\Omega}$. In particular, for a given
continuous boundary value $g$ on $\partial\Omega$, a viscosity solution $u$ of (4.10) in $C(\overline{\Omega})$
with $u = g$ on $\partial\Omega$ is unique.*

*Proof.* We shall prove that $u \leq v$ in $\overline{\Omega}$. Since $\overline{\Omega}$ is compact, by continuity, $u$ and
$v$ are bounded (by Weierstrass' theorem). By adding a suitable constant, we may
assume that $u$ and $v$ are nonnegative, i.e., $u, v \geq 0$ in $\overline{\Omega}$.

It suffices to prove that $\lambda u \leq v$ in $\overline{\Omega}$ for all $\lambda \in (0, 1)$ since $\lim_{\lambda \uparrow 1} \lambda u = u$ in $\overline{\Omega}$.
Note that $u_\lambda = \lambda u$ is a viscosity solution of

$$|\nabla u| - \lambda f(x) = 0 \quad \text{in} \quad \Omega. \tag{4.11}$$

We shall fix $\lambda$ in the sequel.

Although it is logically unnecessary, we first prove that $u_\lambda \leq v$ in $\overline{\Omega}$ when $v \in$
$C^1(\Omega)$ because it reveals the merit of using $u_\lambda$ instead of $u$. If $u_\lambda \leq v$ were false,
then the function $u_\lambda - v$ would take a positive maximum at some $x_* \in \overline{\Omega}$. (The
existence of a maximum follows from Weierstrass' theorem since $\overline{\Omega}$ is compact.) On
the boundary $\partial\Omega$, we know $u_\lambda \leq v$, so $x_* \in \Omega$. Since $u_\lambda$ is a viscosity subsolution
of (4.11), by definition,

$$|\nabla v(x_*)| - \lambda f(x_*) \leq 0.$$

Since $v$ is a classical subsolution of (4.10), we see that

$$|\nabla v(x_*)| - f(x_*) \geq 0.$$

Subtracting the second inequality from the first, we end up with $-\lambda f(x_*) + f(x_*) \leq$
0, which yields a contradiction since $\lambda < 1$ and $f > 0$ on $\Omega$. Unfortunately, this
argument does not work if $v$ is not $C^1$.

To overcome this difficulty, we introduce a doubling-variables method (which
is, of course, different from Kružkov's for conservation law). We note that if $\alpha$ is
large, then $-\Phi_\alpha$ is sufficiently large, i.e., $\Phi_\alpha \ll 0$ away from the diagonal set
$\{(x, x) \mid x \in \overline{\Omega}\}$. We consider

$$\Phi_\alpha(x, y) = u_\lambda(x) - v(y) - \alpha|x - y|^2$$

for a large positive number $\alpha > 0$. Assume that $u_\lambda \leq v$ in $\overline{\Omega}$ would be false. Since we assume $u \geq 0$, we see that $u_\lambda \leq v$ on $\partial\Omega$. Thus, there would exist $x_0 \in \Omega$ such that $m = \Phi_\alpha(x_0, x_0) > 0$. This would imply

$$\max_{\overline{\Omega}\times\overline{\Omega}} \Phi_\alpha \geq m > 0.$$

Let $(x_\alpha, y_\alpha) \in \overline{\Omega}\times\overline{\Omega}$ be a maximizer of $\Phi_\alpha$ over $\overline{\Omega}\times\overline{\Omega}$, i.e., $\max \Phi_\alpha = \Phi_\alpha(x_\alpha, y_\alpha)$. Such $(x_\alpha, y_\alpha)$ exists because of Weierstrass' theorem. Since $m > 0$ and both $u$ and $v$ are bounded as $\alpha \to \infty$, it is easy to see that $\alpha|x_\alpha - y_\alpha|^2$ is bounded. In particular, $x_\alpha - y_\alpha \to 0$ as $\alpha \to \infty$. Since $\Omega$ is bounded so that $\{x_\alpha\}$ is bounded, by compactness (Bolzano–Weierstrass theorem), there is a subsequence $\{x_{\alpha'}\}$ of $\{x_\alpha\}$ converging to some $\hat{x} \in \overline{\Omega}$. Similarly, $\{y_{\alpha'}\}$ has a subsequence $\{y_{\alpha''}\}$ converging to some $\hat{y} \in \overline{\Omega}$. Since $x_\alpha - y_\alpha \to 0$, we see that $\hat{x} = \hat{y}$. We shall denote $\{x_{\alpha''}\}, \{y_{\alpha''}\}$ by $\{x_{\alpha'}\}, \{y_{\alpha'}\}$ for simplicity.

Since we have assumed that $u_\lambda \leq v$ on $\partial\Omega$, we see that $\hat{x} \notin \partial\Omega$. In fact, if $x_{\alpha'}, y_{\alpha'} \to \hat{x} \in \partial\Omega$, then, by the continuity of $u$ and $v$, we see

$$m \leq \limsup_{\alpha'\to\infty} \Phi_{\alpha'}(x_{\alpha'}, y_{\alpha'}) \leq \limsup_{\alpha'\to\infty} (u_\lambda(x_{\alpha'}) - v(y_{\alpha'})) = u_\lambda(\hat{x}) - v(\hat{x}) \leq 0,$$

which is a contradiction (Fig. 4.2).

We take $\alpha$ sufficiently large so that $x_\alpha, y_\alpha \in \Omega$. Since $\Phi$ is maximized at $x_\alpha, y_\alpha$, we see that the function

$$x \mapsto u_\lambda(x) - \varphi_\alpha(x), \qquad \varphi_\alpha(x) = v(y_\alpha) + \alpha|x - y_\alpha|^2$$

takes its maximum at $x_\alpha$ and the function

$$y \mapsto v(y) - \psi_\alpha(y), \qquad \psi_\alpha(y) = u_\lambda(x_\alpha) - \alpha|x_\alpha - y|^2$$

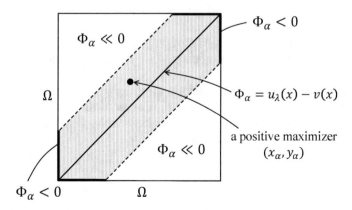

**Fig. 4.2** Values of $\Phi_\alpha$

takes its minimum at $y_\alpha$. By the definition of viscosity sub- and supersolutions, we conclude that

$$|\nabla \varphi_\alpha(x_\alpha)| - \lambda f(x_\alpha) \leq 0,$$

$$|\nabla \psi_\alpha(y_\alpha)| - f(y_\alpha) \geq 0.$$

Subtracting the second inequality from the first and observing that $\nabla_x \varphi_\alpha(x_\alpha) = \nabla_y \psi_\alpha(y_\alpha)$, we now obtain

$$-\lambda f(x_\alpha) + f(y_\alpha) \leq 0.$$

Since $x_{\alpha'} \to \hat{x}$ and $y_{\alpha'} \to \hat{x}$, sending $\alpha' \to \infty$ yields

$$-\lambda f(\hat{x}) + f(\hat{x}) \leq 0.$$

If $f > 0$ on $\Omega$, this leads to a contradiction since $\lambda < 1$. We thus conclude that $\lambda u \leq v$ for all $\lambda \in (0, 1)$, which implies $u \leq v$ in $\overline{\Omega}$.

Suppose that there are two solutions, $u_1$ and $u_2 \in C(\overline{\Omega})$, of (4.10) with $u_1 = u_2 = g$ on $\partial\Omega$. By the comparison just proved, we observe that $u_1 \leq u_2$ and $u_2 \leq u_1$ in $\overline{\Omega}$. This implies $u_1 = u_2$. The proof is now complete. □

The assumption $f(x) > 0$ for all $x \in \Omega$ is essential. If $f$ takes a zero at some point of $\Omega$, the uniqueness actually fails. In fact, if one considers

$$\begin{cases} \left|\dfrac{du}{dx}\right| - |x| = 0, \ |x| < 1 \\ u(\pm 1) = 0, \end{cases}$$

then

$$v_a(x) = \min\left\{(1 - x^2)/2, a + x^2/2\right\}$$

is a viscosity solution for all $a \in [-1/2, 1/2]$ (Fig. 4.3). It turns out that there is at most one solution if all its values on the set $\{x \mid f(x) = 0\}$ are prescribed; see the last paragraph of Sect. 4.5.1.

Note also that there may be no solution for given boundary data. Indeed, if we consider (4.6) in $(-1, 1)$ with $u(-1) = 0$, $u(1) = 3$, then there is no viscosity solution $u \in C[-1, 1]$ satisfying this boundary value. One must interpret the boundary condition in some weak sense.

**Fig. 4.3** Graphs of $v_a$

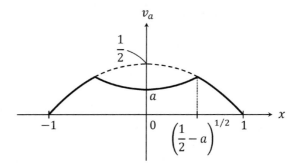

## 4.3 Viscosity Solutions of Evolutionary Hamilton–Jacobi Equations

In this section, we consider an evolutionary Hamilton–Jacobi equation and discuss the uniqueness of viscosity solutions under the periodic boundary condition. The proof is similar to that in the last section.

### 4.3.1 Definition of Viscosity Solutions

We consider an evolutionary Hamilton–Jacobi equation of the form

$$u_t + H(x, \nabla u) = 0 \quad \text{in} \quad \Omega \times (0, T), \tag{4.12}$$

where $\Omega$ is a domain in $\mathbf{R}^N$ or $\mathbf{T}^N$, which imposes a periodic boundary condition. Here we continue to assume that $H$ is a (real-valued) continuous function in $\overline{\Omega} \times \mathbf{R}^N$; $\nabla u$ denotes the (spatial) gradient of a scalar function $u = u(x, t)$ defined on $\Omega \times (0, T)$, i.e., $\nabla u = (\partial_1 u, \ldots, \partial_N u)$.

---

**Definition 4.3**

A function $u \in C(\overline{Q})$ with $Q = \Omega \times (0, T)$ is said to be a *viscosity subsolution* of (4.12) in $Q$ if

$$\varphi_t(\hat{x}, \hat{t}) + H\left(\hat{x}, \nabla\varphi(\hat{x}, \hat{t})\right) \leq 0$$

whenever $\left(\varphi, (\hat{x}, \hat{t})\right) \in C^1(Q) \times Q$ fulfills $\max_Q(u - \varphi) = (u - \varphi)(\hat{x}, \hat{t})$. A function $u \in C(\overline{Q})$ is said to be a *viscosity supersolution* of (4.12) in $Q$ if

$$\varphi_t(\hat{x}, \hat{t}) + H\left(\hat{x}, \nabla\varphi(\hat{x}, t)\right) \geq 0$$

whenever $\left(\varphi, (\hat{x}, \hat{t})\right) \in C^1(Q) \times Q$ fulfills $\min_Q(u - \varphi) = (u - \varphi)(\hat{x}, \hat{t})$. If $u$ is a viscosity sub- and supersolution, we say that $u$ is a *viscosity solution*.

### 4.3.2  Uniqueness

We now present a comparison principle for (4.12) under the periodic boundary condition to simplify the situation. We set $Q_0 = \Omega \times [0, T)$ for later convenience.

> **Theorem 4.4** (Comparison principle)
> *Let $\Omega = \mathbf{T}^N$. Assume that $H(x, p)$ is continuous in $\mathbf{T}^N \times \mathbf{R}^N$. Assume that*
>
> $$|H(x, p) - H(y, p)| \leq \eta\left((1 + |p|)\,|x - y|\right) \quad \text{for all } (x, p) \in \mathbf{T}^N \times \mathbf{R}^N,$$
>
> *where $\eta$ is a modulus, i.e., $\eta(s) > 0$ for $s > 0$ and $\eta(s) \downarrow 0$ as $s \to 0$. Let $u \in C(Q_0)$ and $v \in C(Q_0)$ be viscosity sub- and supersolutions of (4.12), respectively. If $u \leq v$ at $t = 0$, then $u \leq v$ in $Q_0$. In particular, a solution to (4.12) with given initial datum $g \in C(\mathbf{T}^N)$ is unique.*

*Proof.* As in the proof of Theorem 4.2, since the uniqueness (the second statement) easily follows from the comparison principle (the first statement), we just give a proof for the comparison principle. We may assume that $u, v \in C(\overline{Q})$ by taking $T$ smaller. We consider

$$\Phi(x, t, y, s) = u(x, t) - v(y, s) - \alpha|x - y|^2 - \beta|t - s|^2 - \gamma/(T - t) - \gamma/(T - s)$$

for sufficiently large $\alpha, \beta > 0$ and sufficiently small $\gamma > 0$.

Assume that $u \leq v$ in $Q$ were false. Then for sufficiently small $\gamma$, there exists $(x_0, t_0) \in Q$ such that $\Phi(x_0, t_0, x_0, t_0) > 0$. We shall fix such $\gamma$. Then this would imply

$$\max_{\overline{Q} \times \overline{Q}} \Phi = m_{\alpha\beta} > 0.$$

Let $(x_{\alpha\beta}, t_{\alpha\beta}, y_{\alpha\beta}, s_{\alpha\beta}) \in \overline{Q} \times \overline{Q}$ be a maximizer of $\Phi$ over $\overline{Q} \times \overline{Q}$. As in the proof of Theorem 4.2, we see that $\alpha|x_{\alpha\beta} - y_{\alpha\beta}|^2 + \beta|t_{\alpha\beta} - s_{\alpha\beta}|^2$ is bounded as $\alpha \to \infty$, $\beta \to \infty$. In particular, $x_{\alpha\beta} - y_{\alpha\beta} \to 0$, $t_{\alpha\beta} - s_{\alpha\beta} \to 0$ as $\alpha \to \infty$, $\beta \to \infty$.

As in the proof of Theorem 4.2, $t_{\alpha\beta}, s_{\alpha\beta} > 0$ for sufficiently large $\alpha, \beta$ because of the initial condition.

We next observe that

$$\alpha|x_{\alpha\beta} - y_{\alpha\beta}|^2 + \beta|t_{\alpha\beta} - s_{\alpha\beta}|^2 \to 0 \tag{4.13}$$

as $\alpha \to \infty$, $\beta \to \infty$. In fact, since $m_{\alpha\beta} \geq \Phi(x, t, x, t)$, we see that

$$\limsup_{\substack{x-y\to 0 \\ t-s\to 0}} (u(x, t) - v(y, s) - \gamma/(T - t) - \gamma/(T - s)) - m_{\alpha\beta} \leq 0.$$

Setting $(x, t, y, s) = (x_{\alpha\beta}, t_{\alpha\beta}, y_{\alpha\beta}, s_{\alpha\beta})$, we obtain

$$\limsup_{\alpha,\beta\to\infty} \left\{ \Phi(x_{\alpha\beta}, t_{\alpha\beta}, y_{\alpha\beta}, s_{\alpha\beta}) + \alpha|x_{\alpha\beta} - y_{\alpha\beta}|^2 + \beta|t_{\alpha\beta} - s_{\alpha\beta}|^2 - m_{\alpha\beta} \right\} \leq 0,$$

which yields (4.13) since $\Phi(x_{\alpha\beta}, t_{\alpha\beta}, y_{\alpha\beta}, s_{\alpha\beta}) = m_{\alpha\beta}$.

We take $\alpha$, $\beta$ sufficiently large so that $t_{\alpha\beta}, s_{\alpha\beta} > 0$. Since $\Phi$ is maximized at $(x_{\alpha\beta}, t_{\alpha\beta})$, $(y_{\alpha\beta}, s_{\alpha\beta})$, we see that

$$(x, t) \mapsto u(x, t) - \varphi_{\alpha\beta}(x, t),$$

$$\varphi_{\alpha\beta}(x, t) = v(y_{\alpha\beta}, s_{\alpha\beta}) + \alpha|x - y_{\alpha\beta}|^2 + \beta|t - s_{\alpha\beta}|^2 + \gamma/(T - t)$$

takes its maximum at $(x_{\alpha\beta}, t_{\alpha\beta})$. Similarly,

$$(y, s) \mapsto v(y, s) - \psi_{\alpha\beta}(y, s),$$

$$\psi_{\alpha\beta}(y, s) = u(x_{\alpha\beta}, t_{\alpha\beta}) - \alpha|x_{\alpha\beta} - y|^2 - \beta|t_{\alpha\beta} - s|^2 - \gamma/(T - s)$$

takes its minimum at $(y_{\alpha\beta}, s_{\alpha\beta})$. By the definition of viscosity sub- and supersolutions, we conclude that

$$2\beta(t_{\alpha\beta} - s_{\alpha\beta}) + \gamma/(T - t_{\alpha\beta})^2 + H\left(x_{\alpha\beta}, 2\alpha(x_{\alpha\beta} - y_{\alpha\beta})\right) \leq 0,$$

$$2\beta(t_{\alpha\beta} - s_{\alpha\beta}) - \gamma/(T - s_{\alpha\beta})^2 + H\left(y_{\alpha\beta}, 2\alpha(x_{\alpha\beta} - y_{\alpha\beta})\right) \geq 0.$$

Subtracting the second inequality from the first, we conclude that

$$\gamma/(T - t_{\alpha\beta})^2 + \gamma/(T - s_{\alpha\beta})^2 \leq \eta\left((1 + 2\alpha|x_{\alpha\beta} - y_{\alpha\beta}|)\,|x_{\alpha\beta} - y_{\alpha\beta}|\right)$$

by the assumption of continuity of $H$ with respect to $x$. Since $\alpha|x_{\alpha\beta} - y_{\alpha\beta}|^2 \to 0$, $|x_{\alpha\beta} - y_{\alpha\beta}| \to 0$, and $T - t_{\alpha\beta} \leq T$, we conclude that

$$\gamma/T^2 + \gamma/T^2 \leq 0,$$

which yields a contradiction. We thus conclude that $u \leq v$ in $Q_0$.          □

In the proofs of both comparison principles (Theorems 4.2 and 4.4), a key property is that

$$\nabla_x \varphi_\alpha(x_\alpha) = \nabla_y \psi_\alpha(y_\alpha)$$

for Theorem 4.2 and

$$\nabla_x \varphi_{\alpha\beta}(x_{\alpha\beta}, t_{\alpha\beta}) = \nabla_y \psi_{\alpha\beta}(y_{\alpha\beta}, s_{\alpha\beta}), \quad \partial_t \varphi_{\alpha\beta}(x_{\alpha\beta}, t_{\alpha\beta}) = \partial_s \psi_{\alpha\beta}(y_{\alpha\beta}, s_{\alpha\beta})$$

for Theorem 4.4, which follow from

$$\nabla_x |x - y|^2 = -\nabla_y |x - y|^2, \quad \partial_t (t - s)^2 = -\partial_s (t - s)^2.$$

For second derivatives, we have

$$\nabla_x^2 |x - y|^2 = \nabla_y^2 |x - y|^2 \neq -\nabla_y^2 |x - y|^2.$$

This prevents us from extending the foregoing proofs directly to the second-order problems.

## 4.4    Viscosity Solutions with Shock

In this section, we continue to study the uniqueness of a solution for an evolutionary Hamilton–Jacobi equation whose expected solution may develop jump discontinuities called shocks like conservation laws. We first recall the notion of viscosity solutions for semicontinuous functions.

### 4.4.1    Definition of Semicontinuous Functions

We consider an evolutionary Hamilton–Jacobi equation of the form

$$u_t + H(x, t, u, \nabla u) = 0 \quad \text{in} \quad Q = \Omega \times (0, T), \tag{4.14}$$

where $\Omega$ is a domain in $\mathbf{R}^N$ or $\mathbf{T}^N$ and $H$ is a continuous function that may also depend on $t$ and $u$. For a function $u : Q \to \mathbf{R} \cup \{\pm\infty\}$ (i.e., with values in $\mathbf{R} \cup \{\pm\infty\}$), let $u^*$ denote the *upper semicontinuous envelope*, i.e.,

$$u^*(x, t) = \limsup_{\varepsilon \downarrow 0} \left\{ u(y, s) \mid |y - s| < \varepsilon, \ |t - s| < \varepsilon, \ (y, s) \in Q \right\}$$

for $(x, t) \in \overline{Q}$. Similarly, $u_*(x, t)$ denotes the *lower semicontinuous envelope*, i.e., $u_*(x, t) = -(-u)^*(x, t)$ (Exercise 4.3).

**Definition 4.5**

A function $u : Q \to \mathbf{R} \cup \{\pm\infty\}$ is said to be a viscosity subsolution of (4.14) in $Q$ if $u^* < \infty$ on $\overline{Q}$ and

$$\varphi_t(\hat{x}, \hat{t}) + H\left(\hat{x}, \hat{t}, u^*(\hat{x}, \hat{t}), \nabla\varphi(\hat{x}, \hat{t})\right) \leq 0 \tag{4.15}$$

whenever $\left(\varphi, (\hat{x}, \hat{t})\right) \in C^1(Q) \times Q$ fulfills $\max_Q(u^* - \varphi) = (u^* - \varphi)(\hat{x}, \hat{t})$. A viscosity supersolution is defined by replacing $u^*$, $\infty$, $u^*(\hat{x}, \hat{t})$, $\leq$, max by $u_*$, $-\infty$, $u_*(\hat{x}, \hat{t})$, $\geq$, min, respectively. If $u$ is a viscosity sub- and supersolution, we say that $u$ is a viscosity solution.

It is easy to extend Theorem 4.4 to such a discontinuous solution. Moreover, if $r \mapsto H(x, t, r, p)$ is nondecreasing, then $u$ dependence is also allowed.

**Theorem 4.6 (Comparison principle)**
*Assume that $H = H(x, r, p)$ is continuous in $\mathbf{T}^N \times \mathbf{R} \times \mathbf{R}^N$. Assume that $r \mapsto H(x, r, p)$ is nondecreasing and satisfies*

$$\left|H(x, r, p) - H(y, r, p)\right| \leq \eta\big((1 + |p|)\,|x - y|\big), \quad p \in \mathbf{R}^N, \; x, y \in \mathbf{T}^N, \; r \in \mathbf{R}$$

*for some modulus $\eta$. Let $u : Q \to \mathbf{R} \cup \{-\infty\}$ and $v : Q \to \mathbf{R} \cup \{+\infty\}$ be viscosity sub- and supersolutions of (4.14), respectively. If $u^* \leq v_*$ at $t = 0$, then $u^* \leq v_*$ in $Q_0 = \Omega \times [0, T)$. In particular, a solution to (4.14) with $u^*|_{t=0} = u_*|_{t=0} = g \in C(\mathbf{T}^N)$ is unique and continuous in $Q_0$.*

The proof of $u^* \leq v_*$ in $Q_0$ is the same as that of Theorem 4.4, replacing $u$ and $v$ with $u^*$ and $v_*$, respectively, before comparing the inequalities

$$2\beta(t_{\alpha\beta} - s_{\alpha\beta}) + \gamma/(T - t_{\alpha\beta})^2 + H\left(x_{\alpha\beta}, u^*(x_{\alpha\beta}, t_{\alpha\beta}), 2\alpha(x_{\alpha\beta} - y_{\alpha\beta})\right) \leq 0,$$

$$2\beta(t_{\alpha\beta} - s_{\alpha\beta}) - \gamma/(T - s_{\alpha\beta})^2 + H\left(y_{\alpha\beta}, v_*(y_{\alpha\beta}, s_{\alpha\beta}), 2\alpha(x_{\alpha\beta} - y_{\alpha\beta})\right) \geq 0.$$

By the choice of $x_{\alpha\beta}$, $t_{\alpha\beta}$, $y_{\alpha\beta}$, $s_{\alpha\beta}$, we know $u^*(x_{\alpha\beta}, t_{\alpha\beta}) > v_*(y_{\alpha\beta}, s_{\alpha\beta})$. If $r \mapsto H(x, r, p)$ is nondecreasing, we may replace $v_*(y_{\alpha\beta}, s_{\alpha\beta})$ with $u^*(x_{\alpha\beta}, t_{\alpha\beta})$ so that both inequalities are comparable. The remaining part is the same.

If $u_1$ and $u_2$ are solutions with initial datum $g$, the comparison principle implies $u_1^* \leq u_{2*}$ and $u_2^* \leq u_{1*}$. Thus, $u_1 = u_2 \in C(Q_0)$.

We may weaken the monotonicity assumption that $r \mapsto H(x, t, r, p)$ is nondecreasing by a weaker assumption such that $r \mapsto H(x, r, p) + \lambda r$ is nondecreasing for some $\lambda \in \mathbf{R}$ by modifying the structure assumption for $H$. The main idea to extend the proof is the change of dependent variables $u, v$ by $e^{-\lambda t}u$, $e^{-\lambda t}v$. However, if $H$ does not satisfy such monotonicity assumptions, the uniqueness may not hold in general.

## 4.4.2 Example for Nonuniqueness

We consider a scalar conservation law (3.2). Here we assume that $f$ is a given strict convex $C^1$ function in the sense that $f' \in C(\mathbf{R})$ is (strictly) increasing. The equation can be written in the form of (4.14), with

$$H(x, t, r, p) = f'(r)p.$$

For $u_r < u_\ell$, we consider

$$u_s(x, t) = \begin{cases} u_\ell, & x < st, \\ u_r, & x \geq st; \end{cases}$$

see Fig. 4.4. If the speed $s$ satisfies $f'(u_r) \leq s \leq f'(u_\ell)$, then $u_s$ is a viscosity solution in $\mathbf{R} \times (0, \infty)$. (However, it is not a weak solution unless $s$ satisfies the Rankine–Hugoniot condition, i.e., $s = s_*$, with

$$s_* = \frac{f(u_\ell) - f(u_r)}{u_\ell - u_r};$$

see Lemma 3.8.) This shows the nonuniqueness of viscosity solutions. Of course, this is not a direct counterexample of the comparison principle discussed previously since these functions are neither periodic nor continuous up to initial data, but it is easy to construct such an example under the periodic conditions with continuous initial data. In Chap. 3, we introduced the notion of an entropy solution and proved that it was unique. In this example, $u_{s_*}$ is an entropy solution, while $u_s$ with $s \neq s_*$ is not even a weak solution. We shall introduce a notion of a proper solution so that the speed of the jump satisfies the Rankine–Hugoniot condition and entropy condition.

**Fig. 4.4** Graph of $u_s$ at time $t$

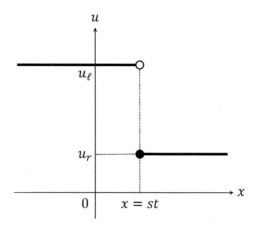

### 4.4.3 Test Surfaces for Shocks

In the definitions of viscosity solutions, we test a possibly irregular function $u$ by a smoother function $\varphi$ (called a test function) from both above and below; see, for example, Definition 4.5. If $u$ is allowed to be discontinuous, as we saw in Sect. 4.4.2, such tests are not enough. To overcome this difficulty, we also test shocks. For simplicity, we consider a one-dimensional setting. In the case where $u$ is discontinuous at $\Gamma$, as in the paragraph after Definition 3.2, but $\Gamma$ may not be smooth, we test the shock $\Gamma$ from both the right and left (or inside or outside with respect to the orientation $\nu^{\ell}$) by a smoother curve called a test curve (Fig. 3.4). The speed of test curves (surfaces) will be given by the Rankine–Hugoniot condition or entropy condition.

For a given point $(x_0, t_0) \in Q$ and $\rho > 0$, $\delta > 0$, let $\{S_t\}_{t \in \Lambda}$ be a smooth family of smooth hypersurfaces in $\mathring{B}_\rho(x_0) \subset \Omega$ with $x_0 \in S_{t_0}$, where $\Lambda = \Lambda_\delta(t_0) = (t_0 - \delta, t_0 + \delta)$, and $B_\rho(x_0)$ denotes a closed ball of radius $\rho$ in $\mathbf{R}^N$ centered at $x_0 \in \mathbf{R}^N$. Let $\mathbf{n} = \mathbf{n}(\cdot, t)$ denote the unit normal vector field of $S_t$ that gives the orientation of $S_t$; we assume that $\mathbf{n}(\cdot, t)$ depends on $t$ at least continuously. Assume that $\mathring{B}_\rho(x_0) \backslash S_t$ consists of two domains. Let $D_t$ denote one of these domains such that $\partial D_t = S_t$ in $\mathring{B}_\rho(x_0)$ and its inward normal agrees with $\mathbf{n} = \mathbf{n}(\cdot, t)$ for $t \in \Lambda_\delta(t_0)$. We call $D_t$ a *region* associated with $(S_t, \mathbf{n}(\cdot, t))$. It is uniquely determined for given $\rho$ and $\delta$.

We simply say that $\{(S_t, \mathbf{n}(\cdot, t))\}$ is an *evolving hypersurface* through $(x_0, t_0)$.

---

**Definition 4.7**

(i) Let $u : Q \to \mathbf{R} \cup \{-\infty\}$ be upper semicontinuous and $(x_0, t_0) \in Q$. For $\mu < u(x_0, t_0)$, we say that an evolving hypersurface $\{(S_t, \mathbf{n}(\cdot, t))\}$ through $(x_0, t_0)$ is an *upper test surface* of $u$ at $(x_0, t_0)$ with level $\mu$ if

$$u(x, t) \leq \mu \quad \text{in} \quad D_t \times \{t\}$$

for some $\rho > 0$ and $\delta > 0$, where $D_t$ denotes the region associated with $(S_t, \mathbf{n}(\cdot, t))$.

(ii) Let $v : Q \to \mathbf{R} \cup \{+\infty\}$ be lower semicontinuous and $(x_0, t_0) \in Q$. For $\mu > v(x_0, t_0)$, we say that an evolving hypersurface $\{(S_t, \mathbf{n}(\cdot, t))\}$ at $(x_0, t_0)$ is a *lower test surface* of $v$ at $(x_0, t_0)$ with level $\mu$ if

$$v(x, t) \geq \mu \quad \text{in} \quad D_t \times \{t\}$$

for some $\rho > 0$, and $\delta > 0$, where $D_t$ denotes the region associated with $(S_t, \mathbf{n}(\cdot, t))$. See Fig. 4.5.

---

If $u(\cdot, t)$ jumps across a hypersurface $\Sigma_t$, such a surface $\Sigma_t$ is often called a shock surface. In this case, one may take $\Sigma_t$ as a test surface if $\Sigma_t$ is regular enough.

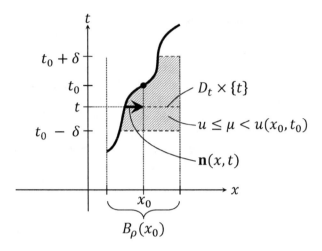

**Fig. 4.5** Upper test surface

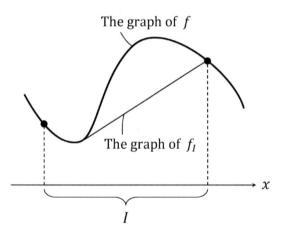

**Fig. 4.6** Convexification

### 4.4.4  Convexification

To give a rigorous definition of solutions, we recall a few properties of convexification. Let $f$ be a function defined on $\mathbf{R}$. Let $I$ be a bounded closed interval. Let $f_I : I \to \mathbf{R}$ denote the *convex hull* (*convexification*) of $f$ in $I$, i.e., $f_I$ is the greatest convex function on $I$ less than or equal to $I$ (Fig. 4.6). By definition, $f_I = f$ in $I$ if $I$ is a singleton.

**Lemma 4.8**

(i) *If $f$ is continuous in $I$, then $f_I = f$ on $\partial I$ and $f_I$ is continuous in $I$.*

(ii) *If $f$ is $C^1$, then $f_I$ is $C^1$ in $I$.*

(iii) *For $f \in C^1[a, d]$ $(-\infty < a < d < \infty)$,*

$$f'_I(x) \geq f'_J(x) \quad \text{for} \quad x \in I \cap J,$$

*with $I = [a, b]$, $J = [c, d]$, $a \leq c \leq b \leq d$, where $'$ denotes the derivative. (At the boundary, the derivative is interpreted as the right or left derivative.)*

(iv) *For $f \in C^1(\mathbf{R})$, the function $F(a, b, x) = f'_{[a,b]}(x)$ is continuous in*

$$\{(p, q, x) \in \mathbf{R}^3 \mid p \leq q, \ p \leq x \leq q\}.$$

The proofs are elementary. They are safely left to the reader; see [46, Lemma 2.1].

## 4.4.5  Proper Solutions

To define a proper solution, we recall a *recession function* of $p = \nabla u$ variable for the Hamiltonian $H : Q \times \mathbf{R} \times \mathbf{R}^N \to \mathbf{R}$, i.e.,

$$H_\infty(x, t, r, p) = \lim_{\lambda \downarrow 0} \lambda H(x, t, r, p/\lambda).$$

See Exercise 4.2. We always assume that $H_\infty$ exists and is continuous in its variables. By definition, $H_\infty(x, t, r, \sigma p) = \sigma H_\infty(x, t, r, p)$ for $\sigma > 0$, i.e., positively homogeneous of degree one in $p$. Indeed,

$$H_\infty(x, t, r, \sigma p) = \lim_{\lambda \downarrow 0} \lambda H\left(x, t, r, \sigma p/\lambda\right) = \lim_{\lambda' \downarrow 0} \sigma \lambda' H\left(x, t, r, p/\lambda'\right).$$

Let $f(r) = f(r; x, t, p)$ be a primitive of $H_\infty(x, t, r, p)$ as a function of $r$. For a closed interval $I$, let $f_I$ denote the convexification of $f$ in $I$. Since $f_I$ is $C^1$ in $I$ by Lemma 4.8 (ii), so that $f'_I$ is continuous on $I$, we set

$$H^I(x, t, r, p) := f'_I(r; x, t, p), \quad r \in I, \quad (x, t) \in Q, \quad p \in \mathbf{R}^N$$

and call $H^I$ a *relaxed Hamiltonian* in $I$. This is independent of the choice of a primitive $f$, so it is well defined. Since $H_\infty$ is positively homogeneous of degree one, so is $H^I$, i.e.,

$$H^I(x, t, r, \sigma p) = \sigma H^I(x, t, r, p)$$

for all $\sigma > 0$, $(x, t) \in Q$, $r \in I$, $p \in \mathbf{R}^N$. If $r \mapsto H(x, t, r, p)$ is nondecreasing so that $f(r)$ is convex, then the relaxed $H^I$ agrees with $H_\infty$ for any choice of $I$.

---

**Definition 4.9**

(i) Let $u : Q \to \mathbf{R} \cup \{-\infty\}$ be a viscosity subsolution of (4.14) in $Q$. We say that $u$ is *proper subsolution* of (4.14) if the inequality

$$V(x_0, t_0) + H^I\big(x_0, t_0, u^*(x_0, t_0), -\mathbf{n}(x_0, t_0)\big) \le 0 \qquad (4.16)$$

holds whenever $(x_0, t_0) \in Q$ admits an upper test surface $\{(S_t, \mathbf{n}(\cdot, t))\}$ of $u^*$ at $(x_0, t_0)$ with the level $\mu (< u^*(x_0, t_0))$, where $I = [\mu, u^*(x_0, t_0)]$. Here, $V = V(x_0, t_0)$ denotes the normal velocity of $\{S_t\}$ at $(x_0, t_0)$ in the direction of $\mathbf{n}(x_0, t_0)$, and $H^I$ denotes the relaxed Hamiltonian.

(ii) For a viscosity supersolution $v : Q \to \mathbf{R} \cup \{+\infty\}$ of (4.14) in $Q$, we say that $v$ is *proper supersolution* of (4.14) if the inequality

$$- V(x_0, t_0) + H^I(x_0, t_0, v_*(x_0, t_0), \mathbf{n}(x_0, t_0)) \ge 0 \qquad (4.17)$$

holds whenever $(x_0, t_0) \in Q$ admits a lower test surface $\{(S_t, \mathbf{n}(\cdot, t))\}$ of $v_*$ at $(x_0, t_0)$ with level $\mu (> v_*(x_0, t_0))$, where $I = [v_*(x_0, t_0), u]$.

(iii) If $u$ is a proper sub- and supersolution, we say that $u$ is a *proper solution*. The notion of proper sub- and supersolution is reduced to classical viscosity sub- and supersolutions respectively if the function is continuous.

▶ **Remark 4.10**

(i) If (4.16) is fulfilled with $I = [\mu, u^*(x_0, t_0)]$, then (4.16) holds for all $I' = [\mu, \sigma]$ provided that $\sigma \ge u^*(x_0, t_0)$ by Lemma 4.8 (iii).

(ii) If $\{(S_t, \mathbf{n}(\cdot, t))\}$ is an upper test surface of $u^*$ at $(x_0, t_0)$ with level $\mu$, then it is also an upper test surface with level $\mu'$ for any $\mu' \in [\mu, u^*(x_0, t_0)]$. Thus, for a proper subsolution, the inequality

$$V(x_0, t_0) + H^J\big(x_0, t_0, u^*(x_0, t_0), -\mathbf{n}(x_0, t_0)\big) \le 0$$

with $J = [\mu', u^*(x_0, t_0)]$ is valid. By Lemma 4.8 (iv), letting $\mu' \uparrow u^*(x_0, t_0)$ yields

$$V(x_0, t_0) + H_\infty\big(x_0, t_0, u^*(x_0, t_0), -\mathbf{n}(x_0, t_0)\big) \le 0, \qquad (4.18)$$

since $H^J = H_\infty$ if $J$ is a singleton. This inequality holds for any upper test surfaces $\{(S_t, \mathbf{n}(\cdot, t))\}$ at $(x_0, t_0)$.

(iii) Suppose that $r \mapsto H(x, t, r, p)$ is nondecreasing so that $H^I = H_\infty$ for any $I$. If (4.18) holds for any upper test surface $\{(S_t, \mathbf{n}(\cdot, t))\}$ at $(x_0, t_0)$ with level $\mu < u^*(x_0, t_0)$, then (4.16) holds for $\mu$ by the monotonicity of $H$ in $r$. Thus, $u$ is a proper subsolution if $u$ is a viscosity subsolution and (4.18) holds for any upper test surface $\{(S_t, \mathbf{n}(\cdot, t))\}$ at $(x_0, t_0)$ with level $\mu < u^*(x_0, t_0)$. In fact, if $r \mapsto H(x, t, r, p)$ is nondecreasing, then every subsolution is a proper subsolution, as stated subsequently in Theorem 4.11.

(iv) For a semiclosed interval $(0, T]$, it is possible to define a proper solution in $Q' = \Omega \times (0, T]$. For $u : Q' \to \mathbf{R} \cup \{\pm\infty\}$, we say that $u$ is a proper subsolution of (4.14) in $Q'$ if it is a viscosity subsolution of (4.14) in $Q'$ (i.e., (4.15) holds for $(\varphi, (\hat{x}, \hat{t})) \in C^1(Q') \times Q'$ satisfying $\max_Q(u^* - \varphi) = (u^* - \varphi)(\hat{x}, \hat{t})$ with $Q$ replaced by $Q'$) and (4.16) holds for upper test surface $\{(S_t, \mathbf{n}(\cdot, t))\}$ at $(x_0, t_0) \in Q'$ with level $\mu \, (< u^*(x_0, t_0))$. If $t_0 = T$, the family $\{(S_t, \mathbf{n}(\cdot, t))\}$ should be interpreted as being smooth in $(T - \delta, T]$.

If $r \mapsto H(x, t, r, p)$ is nondecreasing, then a proper subsolution is a conventional viscosity subsolution under an asymptotic homogeneity assumption on $H$ as $|p| \to \infty$.

**Theorem 4.11 (Consistency)**
*For $H \in C(Q \times \mathbf{R} \times \mathbf{R}^N)$, assume that $r \mapsto H(x, t, r, p)$ is nondecreasing in $\mathbf{R}$ for all $(x, t) \in Q$, $p \in \mathbf{R}^N$. Assume that $\lambda H(x, t, r, p/\lambda)$ converges to $H_\infty$ locally uniformly in $Q \times \mathbf{R} \times \mathbf{R}^N$ as $\lambda \downarrow 0$. In other words,*

$$\lim_{\lambda \downarrow 0} \sup_{(x,t,r,p) \in K} \left| \lambda H\left(x, t, r, \frac{p}{\lambda}\right) - H_\infty(x, t, r, p) \right| = 0 \qquad (4.19)$$

*for every compact set $K$ in $Q \times \mathbf{R} \times \mathbf{R}^N$. If $u$ and $v$ are viscosity sub- and supersolutions of (4.14) in $Q$, then $u$ and $v$ are respectively proper sub- and supersolutions of (4.14) in $Q$.*

▶ **Remark 4.12** By (4.19), the function $H_\infty$ is continuous in its variables. In particular, by the homogeneity of $H_\infty$,

$$H_\infty(x, t, r, 0) = \lim_{\sigma \downarrow 0} H_\infty(x, t, r, \sigma) = \lim_{\sigma \downarrow 0} H_\infty(x, t, r, 1) = 0.$$

By definition,

$$H^I(x, t, r, 0) = 0$$

for any closed interval $I$.

*Proof.* The proof of a viscosity supersolution is similar to that of a viscosity subsolution, so we only present the proof of a viscosity subsolution. By Remark 4.10 (iii), it suffices to prove (4.18). Let $\{(S_t, \mathbf{n}(\cdot, t))\}$ be an upper test surface at $(x_0, t_0) \in Q$ of $u^*$ with level $\mu (< u^*(x_0, t_0))$. Let $D_t$ be a region associated with $(S_t, \mathbf{n}(\cdot, t))$. We set

$$D = \bigcup_{t \in \Lambda} D_t \times \{t\} \subset \mathring{B}_\rho(x_0) \times \Lambda, \qquad \Lambda = (t_0 - \delta, t_0 + \delta).$$

We take another upper test function $\{(S'_t, \mathbf{n}'(\cdot, t))\}$ with level $\mu$ at $(x_0, t_0)$ and $\mathbf{n}(x_0, t_0) = \mathbf{n}'(x_0, t_0)$ such that

$$(x_0, t_0) \in S' \quad \text{and} \quad S' \backslash \{(x_0, t_0)\} \subset D \quad \text{with} \quad S' = \bigcup_{t \in \Lambda} S'_t \times \{t\}.$$

(By construction $S'_t$ touches $S_t$ only at time $t_0$ at point $x_0$.) Let $D'_t$ denote a region associated with $(S'_t, \mathbf{n}'(\cdot, t))$. To construct an appropriate test function[1] for $u^*$, we use a signed distance function of $D' = \bigcup_{t \in \Lambda} D'_t \times \{t\} \subset \mathring{B}_\rho(x_0) \times \Lambda$ defined by

$$d(x, t) = \begin{cases} \operatorname{dist}\big((x, t), \partial D'\big), & x \in D', \\ -\operatorname{dist}\big((x, t), \partial D'\big), & x \notin D'. \end{cases}$$

From this point forward, by $\partial D'$ we mean the boundary of $D'$ in $\mathring{B}_\rho(x_0) \times \Lambda$. Since $\partial D'$ is smooth, so is $d$ in $\mathring{B}_\rho(x_0) \times \Lambda$ for sufficiently small $\delta > 0$ and $\rho > 0$; see, for example, [67]. We fix $\mu' \in (\mu, u^*(x_0, t_0))$ and define

$$\varphi_L(x, t) = \max\big(-Ld(x, t) + u^*(x_0, t_0), \mu'\big)$$

for $L > 0$ (Fig. 4.7). The function $\varphi_L(x, t)$ is smooth outside $D'$ in a small neighborhood of $(x_0, t_0)$. Since $u^*$ is upper semicontinuous, there is a maximizer $(x_L, t_L)$ of $u^* - \varphi_L$ in $B_\rho(x_0) \times \overline{\Lambda}$, where $\overline{\Lambda} = [t_0 - \delta, t_0 + \delta]$. Sending $L \to \infty$ we see that $0 \le \max(u^* - \varphi_L) \to 0$ and $\operatorname{dist}\big((x_L, t_L), \partial D'\big) \to 0$. Since $(x_L, t_L) \notin D'$, this implies $(x_L, t_L) \to (x_0, t_0)$. Moreover, $u^*(x_L, t_L) \to u^*(x_0, t_0)$ since $u^*$ is upper semicontinuous and $u^*(x_L, t_L) \ge u^*(x_0, t_0)$. Thus, for sufficiently large $L$ the function $u^* - \varphi_L$ takes its local maximum at $(x_L, t_L) \in \mathring{B}_\rho(x_0) \times \Lambda$, and at $(x_L, t_L)$ the function $\varphi_L$ is smooth.

---

[1] For a subset $A$ in a metric space $M$ equipped with distance $d$, the distance function $\operatorname{dist}(x, A)$ from $A$ is defined by

$$\operatorname{dist}(x, A) := \inf \big\{ d(x, y) \mid y \in A \big\}.$$

**Fig. 4.7** Graph of $\varphi_L$ at $t = t_0$

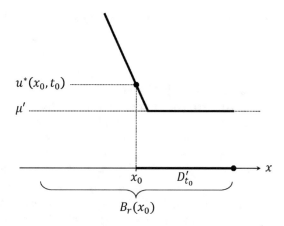

If $u$ is a viscosity subsolution, then

$$\partial_t \varphi_L(x_L, t_L) + H\big(x_L, t_L, u^*(x_L, t_L), \nabla \varphi_L(x_L, t_L)\big) \leq 0.$$

Dividing by $|\nabla \varphi(x_L, t_L)| = L$ and sending $L \to \infty$ yields

$$V + H_\infty\big(x_0, t_0, u^*(x_0, t_0), -\mathbf{n}(x_0, t_0)\big) \leq 0$$

since the convergence $\lambda H(x, t, r, p/\lambda) \to H_\infty(x, t, r, p)$ is locally uniform in $(x, t, r, p)$ as $\lambda \downarrow 0$ and

$$\frac{\nabla \varphi_L(x_L, t_L)}{|\nabla \varphi_L(x_L, t_L)|} \to -\mathbf{n}(x_0, t_0), \qquad (x_L, t_L) \to (x_0, t_0),$$

$$u^*(x_L, t_L) \to u^*(x_0, t_0), \qquad \frac{\partial_t \varphi(x_L, t_L)}{|\nabla \varphi(x_L, t_L)|} \to V(x_0, t_0)$$

$$\text{as} \quad \lambda \downarrow 0.$$

We have thus proved (4.18). □

### 4.4.6 Examples of Viscosity Solutions with Shocks

We consider a scalar conservation law (3.2), where $f$ is a given strict convex $C^1$ function. For $a < b$ we set

$$u_N(x, t) = \begin{cases} a, & x < st, \\ b, & x \geq st, \end{cases}$$

$$u_S(x, t) = \begin{cases} b, & x \leq st, \\ a, & x > st. \end{cases}$$

If the speed $s$ satisfies the Rankine–Hugoniot condition, i.e.,

$$s = \frac{f(b) - f(a)}{b - a},$$

then $u_S$ is a viscosity solution, while $u_N$ is not a viscosity solution even if $s$ satisfies the Rankine–Hugoniot condition. If $s$ satisfies the Rankine–Hugoniot condition, then $u_N$ is still a weak solution (defined in Definition 3.2) of (3.2).

**Proposition 4.13**

*Assume that $f' \in C(\mathbf{R})$ is (strictly) increasing and $a < b$. If $s = (f(b) - f(a))/(b - a)$, then $u_S$ is a proper solution of (3.2). If $s < f'(b)$, then $u_N$ is not even a viscosity supersolution of (3.2). If $s > f'(a)$, then $u_N$ is not even a viscosity subsolution of (3.2).*

*Proof.* It is easy to see that $u_S$ is a viscosity solution. Thus, it suffices to check the speed of a test surface for shocks. Let $(x_0, t_0)$ be a point on a shock, i.e., $x_0 = st_0, t_0 > 0$. The line $S_t = \{x = st\}$ itself is a test surface of $u_S$ and $u_N$ at $(x_0, t_0)$ with level $a$. All other test surfaces at $(x_0, t_0)$ are tangent to $\{S_t\}$, so by Remark 4.10 (ii) it suffices to estimate the normal velocity of $\{S_t\}$. Equation (3.2) can be written

$$u_t + H(u, \nabla u) = 0$$

if we set $H(r, p) = f'(r)p$. If we consider $u_E$, then $\mathbf{n} = 1$, so that

$$H(r, \mathbf{n}) = -f'(r).$$

Since $-f$ is concave,

$$\frac{d}{dr}(-f)_I(r) = -\frac{f(b) - f(a)}{b - a}, \quad r \in I = [a, b],$$

which yields $H^I(r, -1) = -s$ by the definition of $s$. Since $V(x_0, t_0) = c$, we now observe that

$$V(x_0, t_0) + H^I(b, -1) = 0.$$

Thus, $u_S$ is a proper subsolution. A symmetric argument shows that $(u_S)_*$ is a proper supersolution.

It is easy to see that $u_N$ is not a viscosity subsolution or a viscosity supersolution for the range indicated in the statement. □

It is well known (Exercise 3.6) that the entropy solution $u$ with initial datum $u|_{t=0} = u_N|_{t=0}$ is a rarefaction wave solution

$$u_R(x, t) = \begin{cases} a, & x < f'(a)t, \\ (f')^{-1}(x/t), & f'(a)t \le x < f'(b)t, \\ b, & x \ge f'(b)t, \end{cases}$$

where $f'^{-1}$ denotes the inverse function of $f'$. This function $u$ is a continuous viscosity solution, so there are no jumps. Consequently, there are no test surfaces for shocks. Thus, $u_R$ is automatically a proper solution. For $u_R$ and $u_S$, the notions of proper and entropy solutions agree with each other. More generally, it turns out that notions of proper and entropy solutions essentially agree for initial-value problems [46]. We do not touch on this problem in this book.

### 4.4.7 Properties of Graphs

To derive some comparison principle, it is convenient to consider graphs of proper solutions. For a function $u : Q \to \mathbf{R} \cup \{\pm\infty\}$, we associate a function on $\Omega \times \mathbf{R} \times (0, T)$ of the form

$$i_u(x, z, t) = \begin{cases} 0, & z \le u(x, t), \\ -\infty, & z > u(x, t). \end{cases}$$

The set

$$\{i_u = 0\} := \left\{ (x, z, t) \in \Omega \times \mathbf{R} \times (0, T) \mid i_u(x, z, t) = 0 \right\}$$

is called the *subgraph* of $u$ and denoted by $\operatorname{sg} u$. Similarly, we set

$$I_u(x, z, t) = \begin{cases} 0, & z \ge u(x, t), \\ \infty & z < u(x, t). \end{cases}$$

The set

$$\{I_u = 0\} := \left\{ (x, z, t) \in \Omega \times \mathbf{R} \times (0, T) \mid I_u(x, z, t) = 0 \right\}$$

is called the *supergraph* of $u$ and denoted by $\operatorname{Sg} u$. This set $\{I_u = 0\}$ is usually called the epigraph of $u$, and $I_u$ is called the *indicator function* of $\operatorname{Sg} u$ in convex analysis. By definition, $\operatorname{sg} u$ is closed if and only if $u$ is upper semicontinuous. The closure of $\operatorname{sg} u$ equals the subgraph of $u^*$, i.e., $\overline{\operatorname{sg} u} = \operatorname{sg} u^*$. Similarly, $\overline{\operatorname{Sg} u} = \operatorname{Sg} u_*$. By definition, for a function $u : Q \to \mathbf{R} \cup \{\pm\infty\}$, we see that $i_{u^*} = (i_u)^*$, so $i_{u^*}$ is always upper semicontinuous. Similarly, $I_{u_*}$ is always lower semicontinuous.

For later convenience, we first recall the (left) accessibility of a viscosity solution of (4.14).

**Proposition 4.14**
*Assume that $H$ in (4.14) is continuous. Let $u$ be a viscosity subsolution of (4.14) in $Q = \Omega \times (0, T)$. Then $u^*$ is left accessible at each $(x_0, t_0) \in Q$, i.e., there is a sequence $\{(x_j, t_j)\}_{j=1}^{\infty} \subset Q$ such that $x_j \to x_0$, $t_j \uparrow t_0$, $u^*(x_j, t_j) \to u^*(x_0, t_0)$ as $j \to \infty$.*

This follows from the fact that $u$ is a viscosity subsolution of (4.14) in $\Omega \times (0, T']$ for any $T' < T$ and that such a $u$ is left accessible at $t = T'$. We do not give the proof here. For the complete proof, see [22]; see also [47, §3.2.2].

As an application, we obtain some information of functions testing $i_{u^*}$.

**Lemma 4.15**
*Assume the same hypothesis as that of Proposition 4.14. Then $i_{u^*}$ is left accessible in $\hat{Q} = \Omega \times \mathbf{R} \times (0, T)$.*

*Proof.* Assume that $i_{u^*}(x_0, z_0, t_0) = 0$ at $(x_0, z_0, t_0) \in \hat{Q}$, so that $u^*(x_0, t_0) > -\infty$. Since $u^*$ is left accessible at $(x_0, t_0)$ by Proposition 4.14, there is a sequence $\{(x_j, t_j)\}_{j=1}^{\infty} \subset Q$ such that $x_j \to x_0$, $t_j \uparrow t_0$, $u^*(x_j, t_j) \to u^*(x_0, t_0)$ as $j \to \infty$. Since $i_{u^*}(x_0, z_0, t_0) = 0$, we see $z_0 \leq u^*(x_0, t_0)$. If $u^*(x_0, t_0) \in \mathbf{R}$, then we take

$$z_j = u^*(x_j, t_j) - \big(u^*(x_0, t_0) - z_0\big) \leq u^*(x_j, t_j)$$

and observe that $i_{u^*}(x_j, z_j, t_j) = 0$ and $z_j \to z_0$. If $u^*(x_0, t_0) = \infty$, then $z_0 \leq u^*(x_j, t_j)$ for sufficiently large $j$. In this case, we set $z_j = z_0$. We thus conclude that

$$i_{u^*}(x_j, z_j, t_j) = 0, \quad x_j \to x_0, \quad z_j \to x_0, \quad t_j \uparrow t_0$$

as $j \to \infty$. If $i_{u^*}(x_0, z_0, t_0) = -\infty$ so that $z_0 > u^*(x_0, t_0)$, then $z_0 > u^*(x_j, t_j)$ for sufficiently large $j$. Thus, taking $z_j = z_0$, we see that $i_{u^*}(x_j, z_j, t_j) = -\infty$. We now conclude that $i_{u^*}$ is left accessible in $\hat{Q}$.  $\square$

We next check what kind of equations a test function of $i_{u^*}$ satisfies.

**Lemma 4.16**

*Assume that $H$ is continuous and $H_\infty$ exists. Let $u$ be a proper subsolution of (4.14) in $Q = \Omega \times (0, T)$. For $\Phi \in C^1(Q)$ assume that $i_{u*} - \Phi$ takes its maximum over $\hat{Q}$ at $(\hat{x}, \hat{z}, \hat{t})$, i.e.,*

$$\max_{\hat{Q}}(i_{u*} - \Phi) = (i_{u*} - \Phi)(\hat{x}, \hat{z}, \hat{t}).$$

*Then $\partial_z \Phi(\hat{x}, \hat{z}, \hat{t}) \leq 0$ holds.*

(A) *Assume that $\hat{\nabla}\Phi(\hat{x}, \hat{z}, \hat{t}) \neq 0$, where $\hat{\nabla}\Phi = (\nabla_x \Phi, \partial_z \Phi)$. Then $\hat{z} \leq u^*(\hat{x}, \hat{t})$. Moreover,*
   (i) *If $\partial_z \Phi(\hat{x}, \hat{z}, \hat{t}) \neq 0$, then $\hat{z} = u^*(\hat{x}, \hat{t})$ and $\partial_z \Phi(\hat{x}, \hat{z}, \hat{t}) < 0$. Moreover,*

$$\tau + H\left(\hat{x}, \hat{t}, u^*(\hat{x}, \hat{t}), p\right) \leq 0, \tag{4.20}$$

   *with $\tau = -(\partial_t \Phi/\partial_z \Phi)(\hat{x}, \hat{z}, \hat{t}) \in \mathbf{R}$, $p = -(\nabla_x \Phi/\partial_z \Phi)(\hat{x}, \hat{z}, \hat{t}) \in \mathbf{R}^N$.*
   (ii) *If $\partial_z \Phi(\hat{x}, \hat{z}, \hat{t}) = 0$ and $\hat{z} < u^*(\hat{x}, \hat{t})$, then*

$$\partial_t \Phi(\hat{x}, \hat{z}, \hat{t}) + H^I\left(\hat{x}, \hat{t}, u^*(\hat{x}, \hat{t}), \nabla_x \Phi(\hat{x}, \hat{z}, \hat{t})\right) \leq 0, \tag{4.21}$$

   *with $I = [\hat{z}, u^*(\hat{x}, \hat{t})]$.*
   (iii) *Assume (4.19). Assume that $\partial_z \Phi(\hat{x}, \hat{z}, \hat{t}) = 0$ and $\hat{z} = u^*(\hat{x}, \hat{t})$. Then inequality (4.21) holds.*
(B) *Assume (4.19). If $\hat{\nabla}\Phi(\hat{x}, \hat{z}, \hat{t}) = 0$, then $\partial_t \Phi(\hat{x}, \hat{z}, \hat{t}) \leq 0$.*

*Symmetric statements hold for proper supersolutions.*

*Proof.* Since $i_{u*}(x, z, t)$ is nonincreasing in $z$, $(i_{u*} - \Phi)(\hat{x}, z, \hat{t})$ cannot take its maximum at $\hat{z}$ if $\partial_z \Phi(\hat{x}, \hat{z}, \hat{t}) > 0$. Thus, $\partial_z \Phi(\hat{x}, \hat{z}, \hat{t}) \leq 0$.

(A) The point $(\hat{x}, \hat{z}, \hat{t})$ belongs to the boundary of the subgraph $\mathrm{sg}\, u^*$ since $\hat{\nabla}\Phi(\hat{x}, \hat{z}, \hat{t}) \neq 0$. For $\ell = \Phi(\hat{x}, \hat{z}, \hat{t})$ the $\ell$-level set of $\Phi$ touches $\mathrm{sg}\, u^*$ at $(\hat{x}, \hat{z}, \hat{t})$, and the sublevel set $\{\Phi < \ell\}$ does not intersect $\mathrm{sg}\, u^*$. Thus, $\hat{z} \leq u^*(\hat{x}, \hat{t})$. From this point forward, for a function $F$ defined in $\hat{Q}$, by $\{F < \ell\}$ (resp. $\{F \leq \ell\}$) we mean the set

$$\left\{w \in \hat{Q} \mid F(w) < \ell\right\} \quad (\text{resp. } \left\{w \in \hat{Q} \mid F(w) \leq \ell\right\}).$$

(i) By the definition of sg $u^*$, the first statement is clear. Since

$$\partial_z \Phi(\hat{x}, \hat{z}, \hat{t}) < 0,$$

the $\ell$-level set of $\Phi$ can be written as the graph of an implicit function $Z = Z(x, t)$ near $(\hat{x}, \hat{z}, \hat{t})$. By the geometry of the $\ell$-level set of $\Phi$ and sg $u^*$, $u^* - Z$ takes its local maximum at $(\hat{x}, \hat{t})$. Since $Z$ is an implicit function satisfying $\Phi(x, Z(x, t), t) = \ell$, we see $\partial_t Z(\hat{x}, \hat{t}) = \tau$ and $\nabla Z(\hat{x}, \hat{t}) = p$. Since $u^*$ is a subsolution, we get (4.20).

(ii) This is a crucial part of this lemma. We may assume that $\Phi(\hat{x}, \hat{z}, \hat{t}) = 0$ and $\hat{x} = 0$ without loss of generality. Since $\nabla_x \Phi(0, \hat{z}, \hat{t}) \neq 0$, by rotation we may assume that

$$\nabla_x \Phi / |\nabla_x \Phi| = (-1, 0, \dots, 0) \quad \text{at} \quad (0, \hat{z}, \hat{t}).$$

We set

$$\Psi(x, z, t) := (x_1 - R)^2 + x_2^2 + \dots + x_N^2 + (z - \hat{z})^2 + (t - \hat{t})^2 + A(t - \hat{t}) - R^2.$$

For a suitable choice of $R > 0$ and $A \in \mathbf{R}$, a ball $B = \{\Psi \leq 0\}$ touches sg $u^*$ only at $(0, \hat{z}, \hat{t})$ i.e., sg $u^* \cap B = \{(0, \hat{z}, \hat{t})\}$ and $B \subset \{\Phi \leq 0\}$; see Fig. 4.8. Thus,

$$\partial_t \Psi / |\nabla_x \Psi| = \partial_t \Phi / |\nabla_x \Phi| \quad \text{at} \quad (0, \hat{z}, \hat{t}). \tag{4.22}$$

By the choice of $B$ we observe that

$$u(x, t) \leq \hat{z}$$

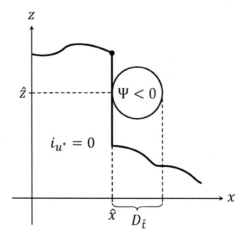

**Fig. 4.8** Ball $\{(x, z) \in \Omega \times \mathbf{R} \mid \Psi(x, z, \hat{t}) < 0\}$

for $x \in D_t = \{x \in \Omega \mid \Psi(x, \hat{z}, t) < 0\}$. We take $S_t$ as the boundary of $D_t$ and $\mathbf{n}$ is the inward normal of $D_t$. By definition, $\{(S_t, \mathbf{n}(\cdot, t))\}$ is an evolving hypersurface through $(0, \hat{t})$, and $D_t$ is a region associated with $\{(S_t, \mathbf{n}(\cdot, t))\}$. Then $\{(S_t, \mathbf{n}(\cdot, t))\}$ is an upper test surface with level $\hat{z}$ of $u^*$ at $(0, \hat{t})$ and $\mathbf{n}(\cdot, \hat{t})$ at $0$ equals $(1, 0, \ldots, 0)$. By (4.22), the normal velocity (in the direction of $\mathbf{n}(\cdot, \hat{t})$) $V$ of $S_{\hat{t}}$ at $\hat{x} = 0$ equals

$$V = \partial_t \Psi / |\nabla_x \Psi| = \partial_t \Phi / |\nabla_x \Phi| \quad \text{at} \quad (0, \hat{z}, \hat{t}).$$

By the definition of a proper subsolution, we see that

$$V + H^I\left(0, \hat{t}, u^*(0, \hat{t}), -\mathbf{n}\right) \le 0, \tag{4.23}$$

with $I = \left[\hat{z}, u^*(0, \hat{t})\right]$. Since

$$V = \frac{\partial_t \Phi}{|\nabla_x \Phi|}(0, \hat{z}, \hat{t}), \quad \mathbf{n} = -\frac{\nabla_x \Phi}{|\nabla_x \Phi|}(0, \hat{z}, \hat{t}),$$

we conclude that (4.23) yields (4.21); here we invoke the homogeneity of $H^I(x, t, r, p)$ in $p$, i.e., $H^I(x, t, r, \sigma p) = \sigma H^I(x, t, r, p)$ for $\sigma > 0$.

(iii) We modify $\Psi$. Let $\tilde{\Psi}$ be a $C^1$ function defined by

$$\tilde{\Psi}(x, z, t) := \begin{cases} \Psi(x, z, t), & \text{if } z \le \hat{z} \\ \Psi(x, z, t) - (z - \hat{z})^2, & \text{if } z > \hat{z}, \end{cases}$$

so that $\partial_z \tilde{\Psi} \le 0$. Since sg $u^*$ is a subgraph, the set $\{\tilde{\Psi} \le 0\}$ still touches sg $u^*$ only at $(0, \hat{z}, \hat{t})$. For $\varepsilon > 0$ we set

$$\Psi_\varepsilon(x, z, t) = \tilde{\Psi}(x, z, t) - \varepsilon(z - \hat{z}).$$

Let $(x_\varepsilon, z_\varepsilon, t_\varepsilon)$ be a maximizer of $i_{u^*} - \Psi_\varepsilon$. Since $i_{u^*} - \tilde{\Psi}$ takes a strict maximum at $(0, \hat{z}, \hat{t})$, by a convergence of maximum points (e.g., [47, Lemma 2.2.5] and Exercise 4.4) $(x_\varepsilon, z_\varepsilon, t_\varepsilon) \to (0, \hat{z}, \hat{t})$ as $\varepsilon \to 0$. Since

$$\partial_z \Psi_\varepsilon(x, z, t) = 2\min(z - \hat{z}, 0) - \varepsilon < 0,$$

we apply (i) to get $z_\varepsilon = u^*(x_\varepsilon, t_\varepsilon)$ and

$$\partial_t \Psi_\varepsilon(x_\varepsilon, z_\varepsilon, t_\varepsilon) + \lambda_\varepsilon H\left(x_\varepsilon, z_\varepsilon, t_\varepsilon, \frac{\nabla_x \Psi_\varepsilon(x_\varepsilon, z_\varepsilon, t_\varepsilon)}{\lambda_\varepsilon}\right) \le 0,$$

with $\lambda_\varepsilon = -\partial_z \Psi_\varepsilon(x_\varepsilon, z_\varepsilon, t_\varepsilon)$ for small $\varepsilon > 0$. By (4.19), letting $\varepsilon \to 0$ yields

$$\partial_t \Psi(0, \hat{z}, \hat{t}) + H_\infty\left(0, \hat{t}, \hat{z}, \nabla_x \Psi(0, \hat{z}, \hat{t})\right) \leq 0$$

since $\lambda_\varepsilon \downarrow 0$. The desired inequality follows from (4.22) and $\nabla_x \Psi / |\nabla_x \Psi| = \nabla_x \Phi / |\nabla_x \Phi|$ at $(0, \hat{z}, \hat{t})$ since $H_\infty$ is positively homogeneous in the variable $\nabla_x \Psi$.

(B) We may assume that $\Phi$ is a separable type of the form

$$\Phi(x, z, t) = \psi(x, z) + a(t), \quad (x, z, t) \in \hat{Q}.$$

We may assume that $i_{u^*} - \Phi$ takes its strict maximum at $(\hat{x}, \hat{z}, \hat{t})$ by replacing $\Phi$ by

$$\Phi + |x - \hat{x}|^2 + (z - \hat{z})^2 + (t - \hat{t})^2.$$

We consider a shift $\Phi_\zeta$ of $\Phi$ by defining

$$\Phi_\zeta(x, z, t) = \Phi(x - \xi, z - \eta, t),$$

with $\zeta = (\xi, \eta) \in \mathbf{R}^N \times \mathbf{R}$. By the convergence of maximum points, there is a sequence $(x_\zeta, z_\zeta, t_\zeta)$ converging to $(\hat{x}, \hat{z}, \hat{t})$ as $\zeta \to 0$ such that

$$\max_{\hat{Q}}(i_{u^*} - \Phi_\zeta) = (i_{u^*} - \Phi_\zeta)(x_\zeta, z_\zeta, t_\zeta).$$

Suppose that there is a sequence $\zeta_j \to 0$ such that

$$\hat{\nabla}\Phi_{\zeta_j}(x_{\zeta_j}, z_{\zeta_j}, t_{\zeta_j}) \neq 0.$$

Since $(x_{\zeta_j}, z_{\zeta_j}, t_{\zeta_j}) \to (\hat{x}, \hat{z}, \hat{t})$, we see

$$\lim_{j \to \infty} \hat{\nabla}\Phi_{\zeta_j}(x_{\zeta_j}, z_{\zeta_j}, t_{\zeta_j}) = \hat{\nabla}\Phi(\hat{x}, \hat{z}, \hat{t}) = 0.$$

If there is a subsequence $\zeta_{j_k}$ such that

$$\partial_z \Phi_{\zeta_{j_k}}(x_{\zeta_{j_k}}, z_{\zeta_{j_k}}, t_{\zeta_{j_k}}) < 0,$$

we apply (A) (i) with $\Phi_{\zeta_{j_k}}$ at $(x_{\zeta_{j_k}}, z_{\zeta_{j_k}}, t_{\zeta_{j_k}})$ to get $z_{\zeta_j} = u^*(x_{\zeta_j}, t_{\zeta_j})$ and

$$\tau_k + H(x_{\zeta_{j_k}}, t_{\zeta_{j_k}}, z_{\zeta_{j_k}}, p_k) \leq 0,$$

with

$$\tau_k = \partial_t \Phi_{\zeta_{j_k}} (x_{\zeta_{j_k}}, z_{\zeta_{j_k}}, t_{\zeta_{j_k}}) \big/ \lambda_k,$$

$$p_k = \nabla_x \Phi_{\zeta_{j_k}} (x_{\zeta_{j_k}}, z_{\zeta_{j_k}}, t_{\zeta_{j_k}}) \big/ \lambda_k,$$

$$\lambda_k = -\partial_z \Phi_{\zeta_{j_k}} (x_{\zeta_{j_k}}, z_{\zeta_{j_k}}, t_{\zeta_{j_k}}) \ (> 0).$$

In other words,

$$\partial_t \Phi_{\zeta_{j_k}} (x_{\zeta_{j_k}}, z_{\zeta_{j_k}}, t_{\zeta_{j_k}})$$
$$+ \lambda_k H \left( x_{\zeta_{j_k}}, t_{\zeta_{j_k}}, z_{\zeta_{j_k}}, \nabla_x \Phi_{\zeta_{j_k}} (x_{\zeta_{j_k}}, z_{\zeta_{j_k}}, t_{\zeta_{j_k}}) \big/ \lambda_k \right) \leq 0.$$

By (4.19), sending $k \to \infty$, we obtain

$$\partial_t \Phi(\hat{x}, \hat{z}, \hat{t}) + H_\infty \left( \hat{x}, \hat{t}, \hat{z}, \nabla_x \Phi(\hat{x}, \hat{z}, \hat{t}) \right) \leq 0$$

since

$$\partial_t \Phi(\hat{x}, \hat{z}, \hat{t}) = \lim_{j \to \infty} \partial_t \Phi_{\zeta_j} (x_{\zeta_j}, z_{\zeta_j}, t_{\zeta_j}),$$

$$\nabla_x \Phi(\hat{x}, \hat{z}, \hat{t}) = \lim_{j \to \infty} \nabla_x \Phi_{\zeta_j} (x_{\zeta_j}, z_{\zeta_j}, t_{\zeta_j}) = 0,$$

$$\partial_z \Phi(\hat{x}, \hat{z}, \hat{t}) = \lim_{j \to \infty} \partial_z \Phi_{\zeta_j} (x_{\zeta_j}, z_{\zeta_j}, t_{\zeta_j}) = 0.$$

Since $H_\infty(\hat{x}, \hat{t}, \hat{z}, 0) = 0$ by Remark 4.12, we now conclude that $\partial_t \Phi(\hat{x}, \hat{z}, \hat{t}) \leq 0$.

If $\partial_z \Phi_{\zeta_j} (x_{\zeta_j}, z_{\zeta_j}, t_{\zeta_j}) = 0$ for sufficiently large $j$, we apply (A) (ii) and (iii) to conclude that $\partial_t \Phi(\hat{x}, \hat{z}, \hat{t}) \leq 0$ since $H^I(\hat{x}, \hat{t}, \hat{z}, 0) = 0$ by Remark 4.12.

It remains to discuss the case where

$$\hat{\nabla} \Phi_\zeta (x_\zeta, z_\zeta, t_\zeta) = 0$$

for sufficiently small $\zeta$. We shall prove that $\Phi$ is independent of $x$ and $z$ near $(\hat{x}, \hat{z}, \hat{t})$. In other words, $\psi$ is constant near $(\hat{x}, \hat{z})$. If so, we are able to conclude that

$$\partial_t \Phi(\hat{x}, \hat{z}, \hat{t}) = \partial_t a(\hat{t}) \leq 0$$

since otherwise it would contradict the left accessibility of $i_{u*}$ in Lemma 4.15.

To show that $\Phi$ is spatially constant near $(\hat{x}, \hat{z}, \hat{t})$, we invoke the following constancy lemma; see [43, Lemma 7.5], where $C^2$ regularity of $\phi$ is assumed. This lemma is implicitly used in [21].

**Lemma 4.17**

*Let $K$ be a compact set in $\mathbf{R}^m$ ($m \geq 2$), and let $h$ be a real-valued upper semicontinuous function on $K$. Let $\phi$ be a $C^1$ function on $\mathbf{R}^d$ with $1 \leq d < m$. Let $G$ be a bounded domain in $\mathbf{R}^d$. For each $\zeta \in G$ assume that there is a maximizer $(r_\zeta, \rho_\zeta) \in K$ of*

$$H_\zeta(r, \rho) = h(r, \rho) - \phi(r - \zeta)$$

*over $K$ such that $\nabla\phi(r_\zeta - \zeta) = 0$. Then*

$$h_\phi(\zeta) = \sup\big\{H_\zeta(r, \rho) \mid (r, \rho) \in K\big\}$$

*is constant in $G$.*

We set $m = N + 2$, $d = N + 1$, and

$$K = \big\{(x, z, t) \in \mathbf{R}^m \mid |x - \hat{x}| + |z - \hat{z}| + |t - \hat{t}| \leq \delta, \ i_{u^*}(x, z, t) = 0\big\} \subset \hat{Q}$$

for some $\delta > 0$. We take $\varepsilon > 0$ small so that $|\zeta| < \varepsilon$ implies

$$\hat{\nabla}\Phi_\zeta(x_\zeta, z_\zeta, t_\zeta) = 0.$$

We then set

$$G = \big\{\zeta \in \mathbf{R}^d \mid |\zeta| < \varepsilon\big\}$$

and take

$$h(r, \rho) = i_{u^*}(r, \rho) - a(\rho) = -a(\rho) \quad \text{on} \quad K$$

$$\phi(r) = \psi(r) \quad \text{on} \quad \mathbf{R}^{N+1},$$

with $r = (x, z)$, $\rho = t$. Here, we extend $\psi$ outside $\Omega \times (0, T)$ so that the extended function is $C^1$ in $\mathbf{R}^{N+1}$. Since $(r_\zeta, \rho_\zeta) = (x_\zeta, z_\zeta, t_\zeta)$ is a minimizer of

$$H_\zeta(r, \rho) = h(r, \rho) - \phi(r - \zeta)$$

over $K$, with $\nabla\phi(r_\zeta - \zeta) = 0$, applying Lemma 4.17 implies that

$$h_\phi(\zeta) = \sup\big\{H_\zeta(r, \rho) \mid (r, \rho) \in K\big\}$$

is constant in $G$. This implies that $\psi$ is constant for $r$ such that $|r - \hat{r}| < \varepsilon$, i.e., $|x - \hat{x}|^2 + |z - \hat{z}|^2 < \varepsilon$. The proof of Lemma 4.16 is now complete.  □

**Proof of Lemma 4.17**  By definition,

$$H_\zeta(r_\eta, \rho_\eta) \le h_\phi(\zeta) = h(r_\zeta, \rho_\zeta) - \phi(r_\zeta - \zeta) \quad \text{for} \quad \zeta, \eta \in G.$$

Since

$$h_\phi(\eta) = H_\eta(r_\eta, \rho_\eta) = h(r_\eta, \rho_\eta) - \phi(r_\eta - \eta) = H_\zeta(r_\eta, \rho_\eta) + \phi(r_\eta - \zeta) - \phi(r_\eta - \eta),$$

we observe that

$$h_\phi(\eta) \le h_\phi(\zeta) + \phi(r_\eta - \zeta) - \phi(r_\eta - \eta).$$

Since $\nabla\phi(r_\eta - \eta) = 0$ and $\phi$ is $C^1$,

$$\left|\phi(r_\eta - \eta) - \phi(r_\eta - \zeta)\right| \le \omega(|\eta - \zeta|)|\eta - \zeta|$$

with some modulus $\omega$, i.e., $\omega(0) = 0$, $\omega \ge 0$, $\omega(\sigma) \to 0$ as $\sigma \to 0$. Here, $\omega$ can be taken to be independent of $\eta$ since $\nabla\phi$ is uniformly continuous on any bounded set. We thus observe that

$$h_\phi(\eta) - h_\phi(\zeta) \le \omega(|\eta - \zeta|)|\eta - \zeta|.$$

Changing the role of $\eta$ and $\zeta$, we end up with

$$\left|h_\phi(\eta) - h_\phi(\zeta)\right| \le \omega(|\eta - \zeta|)|\eta - \zeta|$$

for all $\eta, \zeta \in G$. We now conclude that $h_\phi$ is constant on $G$.  □

### 4.4.8  Weak Comparison Principle

As an application of Lemmas 4.15 and 4.16, we present here a version of a comparison principle for periodic functions. Unlike the earlier comparison principle (Theorem 4.4), the following comparison principle does not imply the uniqueness of a solution. We consider (4.14) with $\Omega = \mathbf{T}^N$ and $H$ independently of $x, t$, i.e.,

$$\partial_t u + H(u, \nabla u) = 0 \quad \text{in} \quad Q = \mathbf{T}^N \times (0, T). \tag{4.24}$$

Theorem 4.18 (Weak comparison principle)
*Assume that $H = H(r, p)$ is continuous and for $M > 0$ there exists a constant $C_M$*

$$\left|H(r, p) - H(r', p)\right| \le C_M |r - r'| (|p| + 1)$$

*for $r, r' \in \mathbf{R}$, with $|r|, |r'| \le M$, $p \in \mathbf{R}^N$. Assume that (4.19) holds for every compact set $K$ in $Q \times \mathbf{R} \times \mathbf{R}^N$. Let $u$ and $v$ be bounded proper sub- and supersolutions of (4.24), respectively. If $u^*(x, 0) < v_*(x, 0)$ for all $x \in \mathbf{T}^N$, then $u^* < v_*$ in $Q$.*

The proof is rather involved compared with that of Theorem 4.4. We present here only the idea of the proof.
*The Idea of the Proof.* We may assume that $u = u^*$, $v = v_*$. Instead of considering $u$ and $v$, we consider $i_u$ and $I_v$ defined in Sect. 4.4.7. We consider

$$\Psi(x, z, t, y, w, s) := i_u(x, z, t) - I_v(y, w, s) - \Xi(x, z, t, y, w, s),$$

$$\Xi(x, z, t, y, w, s) := \alpha|x - y|^2 + \alpha|z - w|^2 + \alpha(t - s)^2 + \sigma/(T - t),$$

where $(x, z), (y, w) \in \mathbf{T}^N \times \mathbf{R}$ and $t, s \in (0, T)$; here, $\alpha$ and $\sigma$ are positive parameters. We argue by contradiction. We fix $\sigma > 0$. We argue in the same way as in the proof of Theorem 4.4 and conclude that a maximizer of $\Psi$ is away from $t = 0$, $s = 0$ for sufficiently large $\alpha$ since initially $i_u(x, z, 0) \le I_v(x, z, 0)$, $(x, z) \in \mathbf{T}^N \times \mathbf{R}$. We divide the situation into two cases.

Case 1.    There is a sequence $\alpha_j \to \infty$ such that at a maximum of $\Psi$ in $\left(\mathbf{T}^N \times \mathbf{R} \times (0, T)\right)^2$ the gradient $(\nabla_x \Xi, \partial_z \Xi) = 0$.
Case 2.    For sufficiently large $\alpha$, $(\nabla_x \Xi, \partial_z \Xi) \ne 0$ at a maximizer of $\Psi$.

In the first case, one gets a contradiction by Lemma 4.16 (B). The second case is itself further subdivided into two cases.

Case 2A.    For sufficiently large $\alpha$, $\partial_z \Xi \ne 0$ at a maximizer of $\Psi$.
Case 2B.    There is a sequence $\alpha_j \to \infty$ such that $\partial_z \Xi = 0$ at a maximizer of $\Psi$.

To derive a contradiction in Case 2A, we use Lemma 4.16 (A) (i).
In Case 2B, we invoke the property of proper solutions. We provide a detailed proof in this case. Let $(\hat{x}, \hat{z}, \hat{t}, \hat{y}, \hat{w}, \hat{s})$ be a maximizer of $\Psi$, with $\hat{t}, \hat{s} > 0$. We have $\partial_z \Xi(\hat{x}, \hat{z}, \hat{t}, \hat{y}, \hat{w}, \hat{s}) = 0$, so that $\hat{z}$ must agree with $\hat{w}$. By Lemma 4.16, $\hat{z} \le u(\hat{x}, \hat{t})$, $\hat{w} \ge v(\hat{y}, \hat{s})$, so that $v(\hat{y}, \hat{s}) \le u(\hat{x}, \hat{t})$.
We shall fix $\alpha$ so that $\hat{t}, \hat{s} > 0$. We first note that

$$a_0 = (\hat{x}, \hat{v}, \hat{t}, \hat{y}, \hat{v}, \hat{s}) \quad \text{and} \quad a_1 = (\hat{x}, \hat{u}, \hat{t}, \hat{y}, \hat{u}, \hat{s}),$$

with

$$\hat{u} = u(\hat{x}, \hat{t}), \quad \hat{v} = v(\hat{y}, \hat{s}),$$

are also maximizers of $\Psi$. Indeed, since $i_u(\hat{x}, z, \hat{t}) = 0$ for all $z \leq \hat{u}$ and $I_v(\hat{y}, w, \hat{s}) = 0$ for all $w \geq \hat{v}$, $\Psi$ must take the same value for $z, w \in \mathbf{R}$ satisfying $z \leq \hat{u}, w \geq \hat{v}$, and $z = w$. In particular, $a_0$ and $a_1$ are maximizers of $\Psi$ since $\hat{v} \leq \hat{u}$.

Since $\Psi$ is maximized at $a_0$, we apply Lemma 4.16 (A) (ii) and (iii) to a function

$$(x, z, t) \mapsto i_u(x, z, t) - \Xi(x, z, t, \hat{y}, \hat{w}, \hat{s}) - I_v(\hat{y}, \hat{w}, \hat{s})$$

to conclude

$$\partial_t \Xi + H^I(\hat{u}, \nabla_x \Xi) \leq 0 \quad \text{at} \quad a_0, \tag{4.25}$$

with $I = [\hat{v}, \hat{u}]$. Similarly, we have

$$-\partial_s \Xi + H^I(\hat{v}, -\nabla_y \Xi) \geq 0 \quad \text{at} \quad a_1. \tag{4.26}$$

Note that $\nabla_x \Xi(a_0) = -\nabla_y \Xi(a_1)$, $\partial_t \Xi(a_0) + \partial_s \Xi(a_1) = \sigma/(T - \hat{t})^2$. Thus, subtracting (4.26) from (4.25) yields

$$\sigma/(T - \hat{t})^2 \leq 0$$

since $H^I(r, p)$ is nondecreasing in $r$ and $\hat{v} \leq \hat{u}$. This yields a contradiction to $\sigma > 0$. This is the end of the idea of the proof.

### 4.4.9  Comparison Principle and Uniqueness

In general, the uniqueness of a solution does not hold even if $H$ is independent of $u$ for discontinuous solutions. As shown in [10], a solution of

$$u_t + (x - t)|u_x| = 0,$$

starting with a characteristic function $1_I$ of some closed interval $I$, is not unique, where $1_I(x) = 1$ for $x \in I$ and $1_I(x) = 0$ for $x \notin I$. This is related to fattening phenomena for a level-set flow of a curvature flow equation; see, for example, [47]. Some additional condition is necessary to guarantee the uniqueness of the initial value problem for (4.24).

**Theorem 4.19** (Strong comparison principle)
*Assume the same hypothesis as that of Theorem 4.18 concerning u, v, and H.*
*Assume furthermore that*

$$- H(r, p) \geq c\sqrt{1 + p^2} \quad with\ some \quad c > 0 \quad for\ all \quad p \in \mathbf{R}^N\ r \in \mathbf{R}.$$

*If $u^*(x, 0) \leq (v_*)^*(x, 0)$ for all $x \in \mathbf{T}^N$, then $u^* \leq (v_*)^*$ in $Q_0 = \mathbf{T}^N \times [0, T)$. If $(u^*)_*(x, 0) \leq v_*(x, 0)$ for all $x \in \mathbf{T}^N$, then $(u^*)_* \leq v_*$ in $Q_0$.*

The proof requires several fundamental properties of viscosity solutions, so we provide only a sketch of the proof.

*Sketch of the Proof.* We provide the proof only where $u^* \leq (v_*)^*$ at $t = 0$ since the proof for the remaining case is symmetric. Again we may assume that $u = u^*$ and $v = v_*$. Since $v$ is a viscosity supersolution of (4.24), it is a viscosity supersolution of

$$w_t - c\sqrt{1 + |\nabla w|^2} = 0$$

by our assumption. This equation has a strong comparison principle (e.g., [10]), so the solution is unique even among semicontinuous functions; see, for example, [51]. The unique upper semicontinuous solution of the $w$ equation with initial datum $w_0$ is given by

$$w(x, t) = \sup \left\{ x \in \mathbf{R} \mid d\left((x, z), \overline{\text{sg}\ w_0}\right) \leq ct \right\},$$

where $\overline{\text{sg}\ w_0}$ denotes the closure of the subgraph sg $w_0$ of $w_0$ defined in Sect. 4.4.7. Heuristically, this is easy to observe since our $w$ equation requires that the graph of $w$ moves with upward normal velocity $V = c$. If one interprets this equation as a surface evolution equation or front propagation of a set $E_0$, then the set $E_t$ at time $t$ is the set of all points whose distance from $E_0$ is less than or equal to $ct$. For more details, see [46]. Since $v$ is a viscosity supersolution of the $w$ equation, the comparison principle for the $w$ equation with initial datum $w_0(x) = v^*(x, 0)$ implies that $v \geq w$ in $\mathbf{T}^N \times (0, T)$. This implies that for $\delta \in (0, T)$ there is $\rho > 0$ that satisfies

$$v(x, t) \geq v^*(x, 0) + \rho \quad \text{for all} \quad x \in \mathbf{T}^N, \quad t \geq \delta. \tag{4.27}$$

We shift $v$ in time and set

$$v_\delta(x, t) = v(x, t + \delta), \quad t > 0.$$

Evidently, $v_\delta$ is a proper supersolution of (4.24) in $\mathbf{T}^N \times (0, T - \delta)$. Assume that $u \le v^*$ at $t = 0$. By (4.27), we see that $u \le v^* \le v_\delta - \rho$ at $t = 0$. Since $v_\delta$ is lower semicontinuous up to $t = 0$, applying weak comparison Theorem 4.18, we obtain $u < v_\delta$ in $\mathbf{T}^N \times [0, T - \delta)$. Sending $\delta$ to zero, we conclude that $u \le v^*$ in $Q_0$ since $\liminf_{\delta \downarrow 0} v_\delta \le v^*$ by the definition of upper semicontinuous function. This is the end of the sketch of the proof.

For a given function $u_0 : \mathbf{T}^N \to \mathbf{R} \cup \{\pm\infty\}$ we say that $u : \mathbf{T}^N \times (0, T) \to \mathbf{R} \cup \{\pm\infty\}$ is a *solution* of (4.24) with initial datum $u_0$ if $u$ is a proper solution of (4.24) and

$$u^*(x, 0) = (u_*)^*(x, 0) = (u_0)^*(x),$$

$$u_*(x, 0) = (u^*)_*(x, 0) = (u_0)_*(x).$$

Our comparison principle implies the uniqueness of a solution.

**Theorem 4.20**
*Assume the same hypothesis as that of Theorem 4.19 concerning H. Let u be a bounded solution of (4.24) with initial datum $u_0$. Then $u^*$ and $u_*$ are unique. Moreover, $(u_*)^* = u^*$, and $(u^*)_* = u_*$.*

*Proof.* Let $v$ be another solution. Since $u^* \le (v_*)^*$ at $t = 0$, the strong comparison principle (Theorem 4.19) implies that $u^* \le (v_*)^* (\le v^*)$ in $\mathbf{T}^N \times (0, T)$. Replacing the roles of $v$ and $u$ yields $v^* \le u^*$. We thus conclude that $v^* = u^*$. Moreover, $u^* \le (u_*)^* \le u^*$ implies $(u_*)^* = u^*$. A symmetric argument implies the uniqueness of $u_*$ and $(u^*)_* = u_*$. □

There are several other situations in which the conclusion of the comparison principle holds. For example, it applies to a conservation law starting with a special class of initial data. The reader is referred to [46] for further examples.

## 4.5    Notes and Comments

### 4.5.1    A Few References on Viscosity Solutions

The theory of viscosity solutions is by now a standard tool to study nonlinear (degenerate) elliptic parabolic partial differential equations of second order as well as first-order equations like Hamilton–Jacobi equations, where expected solutions are not smooth. The notion of a viscosity solution was first introduced by [29] (see also [28]) in a different way for first-order Hamilton–Jacobi equations with a nonconvex Hamiltonian. See the book by Lions [71] for the early stage and the one by Barles [9] for the development of the theory. One of the original applications of the theory is to characterize the value function of control theory and differential games for ordinary differential equations as a unique nondifferentiable solution of Hamilton–Jacobi equations. The reader is referred to the book by Bardi and Capuzzo-Dolcetta [7] as well as [36, Chapter 10].

The extension to second-order equations is not straightforward. It takes several years to overcome the substantial difficulty of obtaining the key comparison principle. The reader is referred to the well-written review article of Crandall, Ishii, and Lions [26] and a shorter review by Ishii [59] for the development of the theory. There are accessible textbooks by Koike [63,64]. The second-order problem relates to stochastic controls. For this type of application see the books by Fleming and Soner [41] and Morimoto [76]. The theory of viscosity solution also gives a mathematical foundation [21, 37] for a level-set flow of the mean curvature flow equations, which was introduced numerically by [81]. For this topic the reader is referred to the book [47], which includes a necessary survey of viscosity solutions. See also the lecture notes of Bardi et al. [8], where various applications, including a level-set method, are presented.

In Sect. 4.2.3, we provide an example of where uniqueness fails for the eikonal equation. For the eikonal equation (4.10), uniqueness with given boundary data is valid provided that the value of a solution on the set $\{x \mid f(x) = 0\}$ is prescribed. This has its origins in the book [71, Section 5.5]. This type of uniqueness and comparison principle is generalized by [38] and [60] for various Hamilton–Jacobi equations with convex Hamiltonians; see also the recent book [87]. This type of comparison principle is roughly stated as follows. If a subsolution $u$ and a supersolution $v$ have an order $u \leq v$ on the (projected) Aubry set $\mathcal{A}$ other than on boundaries, then $u \leq v$ in a whole domain. The Aubry set is a notion related to the Hamilton system corresponding to the Hamilton–Jacobi equation. It consists of an equilibrium set and a point having a sequence of closed curves converging at this point whose Euclidean length is bounded from below, but the corresponding action integral converges to zero. For the eikonal equation (4.10), the Aubry set is simply the equilibrium set $\{x \mid f(x) = 0\}$.

### 4.5.2  Discontinuous Viscosity Solutions

The contents up to Sect. 4.3 are basic materials on viscosity solutions. An elegant proof of the uniqueness of the eikonal equation (Sect. 4.2.3) is due to [58].

Definition 4.5 of the viscosity solution for a semicontinuous function was introduced by [56, 57]. Although this notion is very convenient when it comes to constructing a continuous solution by Perron's method [57], it is not enough to establish the uniqueness of a solution for discontinuous initial data even if the Hamiltonian $H = H(x, t, r, p)$ is independent of $r$, i.e., independent of the value of the unknown function. There are several approaches to recovering uniqueness among semicontinuous functions. When $H = H(x, t, p)$ is convex or concave with respect to $p$, a notion of solutions is introduced by [11] and

[12], so that the solution is unique among semicontinuous functions. For a general $H = H(x, t, p)$, uniqueness was established in [51] based on a level-set method; the main assumption in [51] is that the recession function $H_\infty$ exists.

A proper viscosity solution to handle solution with shock is introduced by [46] to describe a kind of bunching phenomenon of growing crystals on a surface. The contents of Sect. 4.4 are essentially taken from [46]. Since there are many errors in [46], we take this opportunity to correct them. For example, in [46, Proposition 2.5], it was claimed that $u_N$ is a viscosity solution when the speed of a shock comes from the Rankine–Hugoniot condition. However, this statement is wrong. As in Proposition 4.13, this $u_N$ is not even a conditional viscosity solution. Also, "left accessibility" in Proposition 4.14 was written as "right accessibility" in [46]. We also give a detailed proof of Lemma 4.15 (ii) in this book.

There is an interesting way to interpret a proper viscosity solution as an evolution of its graph. If we rewrite the equation for the evolution of the graph, then the graph may not stay as the graph of a function, and the function becomes multivalued. It is natural to think that there is a very singular vertical diffusion that prevents such a phenomenon and causes shocks. This idea is useful for the formulation of a proper viscosity solution [88]. A discussion of the theoretical background of the topic can be found in [44]. There is another approach to interpreting a solution with a shock by introducing an obstacle to prevent overturn [15]. An extension of proper solutions to second-order problems is not yet available.

## 4.6 Exercises

4.1 Find the unique viscosity solution of

$$\begin{cases} \left| \dfrac{du}{dx} \right| - 1 = 0 & \text{in} \quad (-1, 1), \\ u(\pm 1) = 0 \end{cases}$$

and

$$\begin{cases} 1 - \left| \dfrac{du}{dx} \right| = 0 & \text{in} \quad (-1, 1), \\ u(\pm 1) = 0. \end{cases}$$

4.2 For a function $f(p) = \sqrt{1 + |p|^2}$ ($p \in \mathbf{R}^N$) calculate the recession function $f_\infty(p)$.

4.3 Prove that the upper semicontinuous envelope $f^*$ of a real-valued function $f$ in an open set $\Omega \subset \mathbf{R}^N$ is actually upper semicontinuous and that it is smallest among all upper semicontinuous functions greater than or equal to $f$.

4.4 Let $\{f_m\}$ be a sequence of real-valued continuous functions in $\overline{\Omega}$, where $\Omega$ is an open set in $\mathbf{R}^N$. Assume that $f_m$ converges to a continuous function $f$

uniformly in $\overline{\Omega}$ as $m \to \infty$, i.e.,

$$\lim_{m \to \infty} \sup_{x \in \overline{\Omega}} |f_m(x) - f(x)| = 0.$$

Assume that there is $\hat{x}$ such that $f(x) \le f(\hat{x})$ for all $x \in \Omega$ and that $f(x) = f(\hat{x})$ if and only if $x = \hat{x}$. In other words, $f$ takes its strict maximum at $\hat{x}$. Then there is a point $x_m \in \Omega$ that converges to $\hat{x}$ such that $\max_{\overline{\Omega}} f_m = f_m(x_m)$ and $\lim_{m \to \infty} f_m(x_m) = f(\hat{x})$.

4.5 Let $\Omega$ be a domain in $\mathbf{R}^N$. Assume that $u_m \in C(\Omega)$ converges to $u \in C(\Omega)$ locally uniformly in $\Omega$ as $m \to \infty$. Let $u_m$ be a viscosity subsolution of (4.9). Show that $u$ is a viscosity subsolution of (4.9).

4.6 Assume that $f_m \in C([0, 1])$ converges to $f$ uniformly in $[0, 1]$ as $m \to \infty$, i.e.,

$$\lim_{m \to \infty} \sup_{0 \le x \le 1} |f_m(x) - f(x)| = 0.$$

Let $\{x_j\}_{j=1}^{\infty}$ be a sequence in $[0, 1]$ converging to $\hat{x}$ as $j \to \infty$. Show that

$$\lim_{\substack{m \to \infty \\ j \to \infty}} f_m(x_j) = f(\hat{x}).$$

In other words, show that for any $\varepsilon > 0$, there are numbers $m_0$ and $j_0$ such that

$$\left| f_m(x_j) - f(\hat{x}) \right| < \varepsilon$$

for all $m \ge m_0$, $j \ge j_0$.

4.7 Let $P_1$ be the space of all affine functions on $\mathbf{R}^N$, i.e.,

$$P_1 = \left\{ a \cdot x + b \mid a \in \mathbf{R}^N, \ b \in \mathbf{R} \right\}.$$

Let $M$ be a nonempty subset of $P_1$. Set

$$f(x) = \sup \left\{ p(x) \mid p \in M \right\}, \quad x \in \mathbf{R}^N, \tag{4.28}$$

and assume that $f(x)$ is finite. Show that $f$ is a convex function. Show that any real-valued convex function on $\mathbf{R}^N$ is of the form (4.28), with a suitable choice of $M$.

4.8 Let $\Omega$ be a bounded domain in $\mathbf{R}^N$. Let $d$ be the distance function from the boundary $\partial\Omega$, i.e.,

$$d(x) = \inf \left\{ |x - y| \mid y \in \partial\Omega \right\}.$$

Show that $d \in C(\overline{\Omega})$ is the unique viscosity solution of $|\nabla u| = 1$ in $\Omega$ with $u = 0$ on $\partial\Omega$.

4.9 Let $g \in C(\mathbf{R}^N) \cap L^\infty(\mathbf{R}^N)$ be a given function. Show that

$$u(x, t) = \inf \left\{ g(y) + \frac{|x - y|^2}{2t} \,\middle|\, y \in \mathbf{R}^N \right\}$$

is a viscosity solution of

$$v_t + \frac{1}{2}|\nabla v|^2 = 0$$

in $\mathbf{R}^N \times (0, \infty)$. The function $u$ is often called an inf-convolution of $g$.

4.10 Show that

$$u(x, t) = t - |x|, \quad x \in \mathbf{R}, \quad t > 0,$$

is not a viscosity solution of

$$v_t - |\partial_x v| = 0 \tag{4.29}$$

in $\mathbf{R} \times (0, \infty)$, although $u$ satisfies (4.29) outside $x = 0$.

# Appendix: Basic Terminology

**5**

In this appendix, we present definitions of basic terminology used in the book for the reader's convenience. For a given set $W$, $x \in W$ means that $x$ is an *element* of $W$.

## 5.1 Convergence

(1) Let $M$ be a set. A real-valued function $d$ defined on $M \times M$ is said to be a *metric* if
   (i) $d(x, y) = 0$ if and only if $x = y$ for $x, y \in M$;
   (ii) (symmetry) $d(x, y) = d(y, x)$ for all $x, y \in M$;
   (iii) (triangle inequality) $d(x, y) \leq d(x, z) + d(z, y)$ for all $x, y, z \in M$.
   Here, $X_1 \times X_2$ denotes the *Cartesian product* of two sets $X_1$ and $X_2$ defined by

$$X_1 \times X_2 := \left\{ (x_1, x_2) \mid x_i \in X_i \text{ for } i = 1, 2 \right\}.$$

The set $M$ equipped with a metric $d$ is called a *metric space* and denoted by $(M, d)$ if one needs to clarify the metric. Let $W$ be a product of metric spaces of $(M_i, d_i)$ $(i = 1, \ldots, m)$, i.e.,

$$W = \prod_{i=1}^{m} M_i = M_1 \times \cdots \times M_m$$
$$:= \left\{ (x_1, \ldots, x_m) \mid x_i \in M_i \text{ for } i = 1, \ldots, m \right\}.$$

© The Author(s), under exclusive license to Springer Nature Switzerland AG 2023
M.-H. Giga, Y. Giga, *A Basic Guide to Uniqueness Problems for Evolutionary Differential Equations*, Compact Textbooks in Mathematics,
https://doi.org/10.1007/978-3-031-34796-2_5

This $W$ is metrizable, for example, with a metric

$$d(x, y) = \left( \sum_{i=1}^{m} d_i (x_i, y_i)^2 \right)^{1/2}$$

for $x = (x_1, \ldots, x_m)$, $y = (y_1, \ldots, y_m) \in W$. If $M_i$ is independent of $i$, i.e., $M_i = M$, then we simply write $W$ as $M^m$.

A subset $A$ of $M$ is said to be *open* if for any $x \in A$ there is $\varepsilon > 0$ such that the *ball* $B_\varepsilon(x) = \{ y \in M \mid d(y, x) \le \varepsilon \}$ is included in $A$. If the *complement* $A^c$ is open, then $A$ is said to be *closed*. The complement $A^c$ is defined by

$$A^c = M \backslash A := \{ x \in M \mid x \notin A \} .$$

For a set $A$, the smallest closed set including $A$ is called the *closure* of $A$ and denoted by $\overline{A}$. Similarly, the largest open set included in $A$ is called the *interior* of $A$ and denoted by int $A$ or simply by $\overset{\circ}{A}$. By definition, $A = \overline{A}$ if and only if $A$ is closed, and $A = \overset{\circ}{A}$ if and only if $A$ is open. The set $\overline{A} \backslash \overset{\circ}{A}$ is called the *boundary* of $A$ and denoted by $\partial A$. For a subset $B$ of a set $A$, we say that $B$ is *dense* in $A$ if $\overline{B} = A$. A set $A$ in $M$ is *bounded* if there is $x_0 \in M$ and $R > 0$ such that $A$ is included in $B_R(x_0)$. For a *mapping* $f$ from a set $S$ to $M$ (i.e., an $M$-valued function defined on $S$), $f$ is said to be *bounded* if its *image* $f(S)$ is bounded in $M$, where

$$f(S) = \{ f(x) \mid x \in S \} .$$

(2) Let $V$ be a real vector space (a vector space over the field $\mathbf{R}$). A nonnegative function $\| \cdot \|$ on $V$ is said to be a *norm* if
   (i) $\|x\| = 0$ if and only if $x = 0$ for $x \in V$;
   (ii) $\|cx\| = |c| \|x\|$ for all $x \in V$ and all $c \in \mathbf{R}$;
   (iii) (triangle inequality) $\|x + y\| \le \|x\| + \|y\|$ for all $x, y \in V$.
The vector space $V$ equipped with a norm $\| \cdot \|$ is called a *normed vector space* and denoted by $(V, \| \cdot \|)$ if one needs to clarify the norm. By definition,

$$d(x, y) = \|x - y\|$$

is a metric. A normed vector space is regarded as a metric space with the foregoing metric.

(3) Let $\{z_j\}_{j=1}^{\infty}$ be a sequence in a metric space $(M, d)$. We say that $\{z_j\}_{j=1}^{\infty}$ *converges* to $z \in M$ if for any $\varepsilon > 0$ there exists a natural number $n = n(\varepsilon)$ such that $j \ge n(\varepsilon)$ implies $d(z, z_j) < \varepsilon$. In other words,

$$\lim_{j \to 0} d(z, z_j) = 0.$$

We simply write $z_j \to z$ as $j \to \infty$, or $\lim_{j\to\infty} z_j = z$. If $\{z_j\}_{j=1}^\infty$ converges to some element, we say that $\{z_j\}_{j=1}^\infty$ is a *convergent sequence*.

(4) Let $f$ be a mapping from a metric space $(M_1, d_1)$ to another metric space $(M_2, d_2)$. We say that $f(y)$ *converges* to $a \in M_2$ as $y$ tends to $x$ if for any $\varepsilon > 0$ there exists $\delta = \delta(\varepsilon) > 0$ such that

$$d_2(f(y), a) < \varepsilon \quad \text{if} \quad d_1(y, x) < \delta.$$

We simply write $f(y) \to a$ as $y \to x$ or $\lim_{y\to x} f(y) = a$. If

$$\lim_{y\to x} f(y) = f(x),$$

then $f$ is said to be *continuous* at $x \in M_1$. If $f$ is continuous at all $x \in M_1$, then $f$ is said to be *continuous* on $M_1$ (with values in $M_2$). The space of all continuous functions on $M_1$ with values in $M_2$ is denoted by $C(M_1, M_2)$.

(5) Let $\{z_j\}_{j=1}^\infty$ be a sequence in a metric space $(M, d)$. We say that $\{z_j\}_{j=1}^\infty$ is a *Cauchy sequence* if for any $\varepsilon > 0$ there exists a natural number $n = n(\varepsilon)$ such that $j, k \geq n(\varepsilon)$ implies $d(z_j, z_k) < \varepsilon$. It is easy to see that a convergent sequence is always a Cauchy sequence, but the converse may not hold. We say that the metric space $(M, d)$ is *complete* if any Cauchy sequence is a convergent sequence.

(6) Let $(V, \|\cdot\|)$ be a normed vector space. We say that $V$ is a *Banach space* if it is complete as a metric space. The norm $\|\cdot\|$ is often written as $\|\cdot\|_V$ to distinguish it from other norms if we use several norms. We simply write $z_j \to z$ in $V$ (as $j \to \infty$) if $\lim_{j\to\infty} \|z_j - z\|_V = 0$ and $z \in V$ for a sequence $\{z_j\}_{j=1}^\infty$. We often say that $z_j$ converges to $z$ *strongly* in $V$ (as $j \to \infty$) to distinguish this convergence from other weaker convergences discussed later.

(7) Let $V$ be a real vector space. A real-valued function $\langle \cdot, \cdot \rangle$ defined on $V \times V$ is said to be an *inner product* if
  (i) $\langle x, x \rangle \geq 0$ for all $x \in V$;
  (ii) $\langle x, x \rangle = 0$ if and only if $x = 0$;
  (iii) (symmetry) $\langle x, y \rangle = \langle y, x \rangle$ for all $x, y \in V$;
  (iv) (linearity) $\langle c_1 x_1 + c_2 x_2, y \rangle = c_1 \langle x_1, y \rangle + c_2 \langle x_2, y \rangle$ for all $x_1, x_2, y \in V$, $c_1, c_2 \in \mathbf{R}$.
By definition, it is easy to see that

$$\|z\| = \langle z, z \rangle^{1/2}$$

is a norm. The space with an inner product is regarded as a normed vector space with the foregoing norm. If this space is complete as a metric space, we say that $V$ is a *Hilbert space*. The Euclidean space $\mathbf{R}^N$ is a finite-dimensional Hilbert space equipped with a standard inner product. It turns out that any finite-dimensional Hilbert space is "isomorphic" to $\mathbf{R}^N$. Of course, a Hilbert space is an example of a Banach space.

(8) Let $V$ be a Banach space equipped with norm $\| \cdot \|$. Let $V^*$ denote the totality of all continuous linear function(al)s on $V$ with values in $\mathbf{R}$. (By the Hahn–Banach theorem, the vector space $V^*$ has at least one dimension. Incidentally, Mazur's theorem in the proof of Lemma 1.19 in Sect. 1.2.3 is another application of the Hahn–Banach theorem.)

The space $V^*$ is called the *dual space* of $V$. Let $\{z_j\}_{j=1}^{\infty}$ be a sequence in $V^*$. We say that $\{z_j\}_{j=1}^{\infty}$ converges to $z \in V^*$ *-weakly* if

$$\lim_{j \to \infty} z_j(x) = z(x)$$

for any $x \in V$. We often write $z_j \overset{*}{\rightharpoonup} z$ in $V^*$ as $j \to \infty$. Such a sequence $\{z_j\}_{j=1}^{\infty}$ is called a *-weak convergent sequence*. The dual space $V^*$ is equipped with the norm

$$\|z\|_{V^*} := \sup \left\{ z(x) \mid \|x\| = 1, \ x \in V \right\} = \sup_{\|x\|=1} z(x).$$

The space $V^*$ is also a Banach space with this norm. Here, for a subset $A$ in $\mathbf{R}$, by $a = \sup A$ we mean that $a$ is the smallest real member that satisfies $a \geq x$ for any $a \in A$. In other words, it is the least upper bound of $A$. The notation sup is the abbreviation of the *supremum*. Similarly, inf $A$ denotes the greatest lower bound of $A$, and it is the abbreviation of the *infimum*. If $\sup A = a$ with $a \in A$, we write max $A$ instead of sup $A$. The same convention applies to inf and min.

Since $V^*$ is a Banach space, there is a notion of convergence in the metric defined by the norm. To distinguish this convergence from *-weak convergence, we say that $\{z_j\}_{j=1}^{\infty}$ converges to $z$ *strongly* in $V^*$ if

$$\lim_{j \to \infty} \|z_j - z\|_{V^*} = 0,$$

and it is simply written $z_j \to z$ in $V^*$ as $j \to \infty$. By definition, $z_j \to z$ implies $z_j \overset{*}{\rightharpoonup} z$, but the converse may not hold.

(9) Let $A$ be a subset of a metric space $M$. The set $A$ is said to be (sequentially) *relatively compact* if any sequence $\{z_j\}_{j=1}^{\infty}$ in $A$ has a convergent subsequence in $M$. If, moreover, $A$ is closed, we simply say that $A$ is *compact*. When $A$ is compact, it is always bounded. When $A$ is a subset of $\mathbf{R}^N$, it is well known as the Bolzano–Weierstrass theorem that $A$ is compact if and only if $A$ is bounded and closed. However, if $A$ is a subset of a Banach space $V$, such an equivalence holds if and only if $V$ is of finite dimension. In other words, a bounded sequence of an infinite-dimensional Banach space may not have a (strongly) convergent subsequence.

There is a compactness theorem (Banach–Alaoglu theorem) that says if $\{z_j\}_{j=1}^{\infty}$ in a dual Banach space $V^*$ is bounded, i.e.,

$$\sup_{j \geq 1} \|z_j\|_{V^*} < \infty,$$

then it has a $*$-weak convergent subsequence (Exercise 1.9).

(10) Let $V$ be a Banach space and $V^*$ denote its dual space. Let $\{x_k\}_{k=1}^{\infty}$ be a sequence in $V$. We say that $\{x_k\}_{k=1}^{\infty}$ converges to $x \in V$ *weakly* if

$$\lim_{k \to \infty} z(x_k) = z(x)$$

for all $z \in V^*$. We often write $x_k \rightharpoonup x$ in $V$ as $k \to \infty$. Such a sequence is called a *weak convergent sequence*.

If a Banach space $W$ is a dual space of some Banach space $V$, say, $W = V^*$, there are two notions, weak convergence and $*$-weak convergence. Let $\{z_j\}_{j=1}^{\infty}$ be a sequence in $W$. By definition, $z_j \overset{*}{\rightharpoonup} z$ (in $W$ as $j \to \infty$) means that $\lim_{j \to \infty} z_j(x) = z(x)$ for all $x \in V$ while $z_j \rightharpoonup z$ (in $W$ as $j \to \infty$) means that $\lim_{j \to \infty} y(z_j) = y(z)$ for all $y \in W^* = (V^*)^*$.

The space $V$ can be continuously embedded in $V^{**} = (V^*)^*$. However, $V$ may not be equal to $V^{**}$. Thus, weak convergence is stronger than $*$-weak convergence. If $V = V^{**}$, then both notions are the same. The space $V$ is called *reflexive* if $V = V^{**}$.

(11) If $V$ is a Hilbert space, it is reflexive. More precisely, the mapping $x \in V$ to $z \in V^*$ defined by

$$z(y) = \langle x, y \rangle, \quad y \in V$$

is a linear isomorphism from $V$ to $V^*$, which is also norm preserving, i.e., $\|z\|_{V^*} = \|x\|$. This result is known as the Riesz–Fréchet theorem. Thus, the notions of weak convergence and $*$-weak convergence are the same.

(12) Let $f$ be a real-valued function in a metric space $M$. We say that $f$ is *lower semicontinuous* at $x \in M$ if

$$f(x) \leq \liminf_{y \to x} f(y) := \liminf_{\delta \downarrow 0} \left\{ f(y) \mid d(y, x) < \delta \right\},$$

where $\lim_{\delta \downarrow 0}$ denotes the limit as $\delta \to 0$ but restricted to $\delta > 0$. Even if $f$ is allowed to take $+\infty$, the definition of the lower semicontinuity will still be valid. If $f$ is lower semicontinuous for all $x \in M$, we simply say that $f$ is lower semicontinuous on $M$. If $-f$ is lower semicontinuous, we say that $f$ is *upper semicontinuous*.

(13) Let $f = f(t)$ be a function of one variable in an interval $I$ in $\mathbf{R}$ with values in a Banach space $V$. We say that $f$ is *right differentiable* at $t_0 \in I$ if there is $v \in V$ such that

$$\lim_{h \downarrow 0} \| f(t_0 + h) - f(t_0) - vh \| \,/\, h = 0$$

provided that $t_0 + h \in I$ for sufficiently small $h > 0$. Such $v$ is uniquely determined if it exists and is denoted by

$$v = \frac{\mathrm{d}^+ f}{\mathrm{d}t}(t_0).$$

This quantity is called the *right differential* of $f$ at $t_0$. The function $t \mapsto \frac{\mathrm{d}^+ f}{\mathrm{d}t}(t)$ is called the *right derivative* of $f$. The left differentiability is defined in a symmetric way by replacing $h \downarrow 0$ with $h \uparrow 0$. Even if both right and left differentials exist, they may be different. For example, consider $f(t) = |t|$ at $t_0 = 0$. The right differential at zero is 1, while the left differential at zero is $-1$. If the right and left differentials agree with each other at $t = t_0$, we say that $f$ is *differentiable* at $t = t_0$, and its value is denoted by $\frac{\mathrm{d}f}{\mathrm{d}t}(t_0)$. The function $t \mapsto \frac{\mathrm{d}f}{\mathrm{d}t}(t)$ is called the *derivative* of $f$. If $f$ depends on other variables, we write $\partial f / \partial t$ instead of $\mathrm{d}f / \mathrm{d}t$ and call the *partial derivative* of $f$ with respect to $t$.

## 5.2    Measures and Integrals

(1) For a set $M$, let $2^M$ denote the family of all subsets of $H$. We say that a function $\mu$ defined on $2^M$ with values in $[0, \infty]$ is an (outer) *measure* if
  (i) $\mu(\emptyset) = 0$;
  (ii) (countable subadditivity) $\mu(A) \leq \sum_{j=1}^{\infty} \mu(A_j)$ if a countable family $\{A_j\}_{j=1}^{\infty}$ covers $A$, where $A_j, A \in 2^M$. In other words, $A$ is included in a union of $\{A_j\}_{j=1}^{\infty}$, i.e., a point of $A$ must be an element of some $A_j$.
  Here, $\emptyset$ denotes the empty set.
(2) A set $A \in 2^M$ is said to be $\mu$-*measurable* if

$$\mu(S \cap A^c) + \mu(S \cap A) = \mu(S)$$

for any $S \in 2^M$. Let $M_0$ be a metric space. A mapping $f$ from $M$ to $M_0$ is said to be $\mu$-*measurable* if the *preimage* $f^{-1}(U)$ of an open set $U$ of $M_0$ is $\mu$-measurable. Here,

$$f^{-1}(U) := \{ x \in M \mid f(x) \in U \}.$$

A set $A$ with $\mu(A) = 0$ is called a $\mu$-*measure zero set*. If a statement $P(x)$ for $x \in M$ holds for $x \in M \backslash A$ with $\mu(A) = 0$, we say that $P(x)$ holds for $\mu$-*almost every* $x \in M$ or shortly a.e. $x \in M$. In other words, $P$ holds in $M$ outside a $\mu$-measure zero set. In this case, we simply say that $P$ holds *almost everywhere* in $M$.

Let $\mathcal{M}$ be the set of all $\mu$-measurable sets. If we restrict $\mu$ just to $\mathcal{M}$, i.e., $\overline{\mu} = \mu|_{\mathcal{M}}$, then $\overline{\mu}$ becomes a measure on $\mathcal{M}$. Since in this book we consider $\mu(A)$ for a $\mu$-measurable set $A$, we often say simply a measure instead of an outer measure.

(3) Let $A$ be a subset of $\mathbf{R}^N$. Let $\mathcal{C}$ be a family of closed cubes in $\mathbf{R}^N$ whose faces are orthogonal to the $x_i$-axis for some $i = 1, \ldots, N$. In other words, $C \in \mathcal{C}$ means

$$C = \left\{ (x_1, \ldots, x_N) \in \mathbf{R}^N \mid a_i \leq x_i \leq a_i + \ell \ (i = 1, \ldots, n) \right\}$$

for some $a_i, \ell \in \mathbf{R}$. Let $|C|$ denote its volume, i.e., $|C| = \ell^N$. We set

$$\mathcal{L}^N(A) = \inf \left\{ \sum_{j=1}^{\infty} |C_j| \ \middle| \ \{C_j\}_{j=1}^{\infty} \text{ covers } A \text{ with } C_j \in \mathcal{C} \right\}.$$

It turns out that $\mathcal{L}^N(C) = |C|$; it is nontrivial to prove $\mathcal{L}^N(C) \geq |C|$. It is easy to see that $\mathcal{L}^N$ is an (outer) measure in $\mathbf{R}^N$. This measure is called the *Lebesgue measure* in $\mathbf{R}^N$. It can be regarded as a measure in the flat torus $\mathbf{T}^N = \prod_{i=1}^{N} (\mathbf{R}/\omega_i \mathbf{Z})$. For a subset $A$ of $\mathbf{T}^N$, we regard this set as a subset $A_0$ of the *fundamental domain* (i.e., the *periodic cell* $[0, \omega_1) \times \cdots \times [0, \omega_N)$). The Lebesgue measure of $A$ is defined by $\mathcal{L}^N(A) = \mathcal{L}^N(A_0)$. Evidently, $\mathcal{L}^N(\mathbf{T}^N) = \omega_1 \cdots \omega_N$, which is denoted by $|\mathbf{T}^N|$ in the proof of Lemma 1.21 in Sect. 1.2.5.

(4) In this book, we only use the Lebesgue measure. We simply say measurable when a mapping or a set is $\mathcal{L}^N$-measurable. Instead of writing $\mathcal{L}^N$-a.e., we simply write a.e. Let $\Omega$ be a measurable set in $\mathbf{T}^N$ or $\mathbf{R}^N$, for example, $\Omega = \mathbf{T}^N$. Let $f$ be a measurable function on $\Omega$ with values in a Banach space $V$. Then one is able to define its integral over $\Omega$. When $V = \mathbf{R}^m$, this integral is called the Lebesgue integral. In general, it is called the Bochner integral of $f$ over $\Omega$. Its value is denoted by $\int_\Omega f \, d\mathcal{L}^N$ or, simply, $\int_\Omega f \, dx$; See, for example, [90, Chapter V, Section 5]. If $\Omega = \mathbf{T}^N$ and $f$ is continuous, this agrees with the more conventional Riemann integral. For $p \in [1, \infty)$ and a general Banach space $V$, let $\tilde{L}^p(\Omega, V)$ denote the space of all measurable functions $f$ with values in $V$ such that

$$\|f\|_p = \left( \int_\Omega \|f(x)\|^p \, dx \right)^{1/p}$$

is finite. If $\|f\|_p$ is finite, we say that $f$ is $p$th *integrable*. If $p = 1$, we simply say $f$ is integrable. If $f \in \tilde{L}^1(\Omega, V)$, we say that $f$ is *integrable* in $\Omega$. We identify two functions $f, g \in \tilde{L}^p(\Omega, V)$ if $f = g$ a.e. and define $L^p(\Omega, V)$ from $\tilde{L}^p(\Omega, V)$ by this identification. It is a fundamental result that $L^p(\Omega, V)$ is a Banach space equipped with the norm $\|\cdot\|_p$. When $V = \mathbf{R}$, we simply write $L^p(\Omega)$ instead of $L^p(\Omega, V)$. The case $p = \infty$ should be handled separately. For a general Banach space $V$, let $\tilde{L}^\infty(\Omega, V)$ denote the space of all measurable functions $f$ with values in $V$ such that

$$\|f\|_\infty = \inf \left\{ \alpha \ \Big| \ \mathcal{L}^N \left( \{x \in \Omega \mid \|f(x)\|_V > \alpha\} \right) = 0 \right\}$$

is finite. By the same identification, the space $L^\infty(\Omega, V)$ can be defined. This space $L^\infty(\Omega, V)$ is again a Banach space. Key theorems in the theory of Lebesgue integrals used in this book include the Lebesgue dominated convergence theorem and Fubini's theorem. Here, we give a version of the *dominated convergence theorem*.

**Theorem 5.1**
*Let $V$ be a Banach space. Let $\{f_m\}_{m=1}^\infty$ be a sequence in $L^1(\Omega, V)$. Assume that there is a nonnegative function $\varphi \in L^1(\Omega)$ independent of $m$ such that $\|f_m(x)\|_V \le \varphi(x)$ for a.e. $x \in \Omega$. If $\lim_{m\to\infty} f_m(x) = f(x)$ for a.e. $x \in \Omega$, then*

$$\lim_{m\to\infty} \int_\Omega f_m(x)\, dx = \int f(x)\, dx.$$

*In other words, $\lim_{m\to\infty} \left\| \int_\Omega f_m\, dx - \int_\Omega f(x)\, dx \right\|_V = 0$.*

Usually, $V$ is taken as $\mathbf{R}$ or $\mathbf{R}^N$, but it is easy to extend to this setting. For basic properties of the Lebesgue measure and integrals, see for example a classical book of Folland [42]. We take this opportunity to clarify $*$-weak convergence in $L^p$ space. A basic fact is that $(L^p(\Omega))^* = L^{p'}(\Omega)$ for $1 \le p < \infty$, where $1/p + 1/p' = 1$. Note that $p = \infty$ is excluded, but $(L^1)^* = L^\infty$. Since $L^p$ is reflexive for $1 < p < \infty$, weak convergence and $*$-weak convergence agree with each other. Let us write a $*$-weak convergence in $L^\infty$ explicitly. A sequence $\{f_m\}$ in $L^\infty(\Omega)$ $*$-weakly converges to $f \in L^\infty(\Omega)$ as $m \to \infty$ if and only if

$$\lim_{m\to\infty} \int_\Omega f_m \varphi\, dx = \int_\Omega f\varphi\, dx$$

for all $\varphi \in L^1(\Omega)$. For detailed properties of $L^p$ spaces, see, for example, [19, Chapter 4].

For a Banach space $V$-valued $L^p$ function, we also consider its dual space. That is, we have

$$\left(L^p(\Omega, V)\right)^* = L^{p'}(\Omega, V^*) \quad \text{for} \quad 1 \le p < \infty$$

with $1/p + 1/p' = 1$. (This duality—at least for reflexive $V$—can be proved along the same line as in [19, Chapter 4], where $V$ is assumed to be $\mathbf{R}$. For a general Banach space $V$, see, for example, [35, Chapter IV].) We consider $*$-weak convergence in $L^\infty(\Omega, V)$ with $V = L^q(U)$, $1 < q \le \infty$, where $\Omega$ is an open interval $(0, T)$ and $U$ is an open set in $\mathbf{T}^N$ or $\mathbf{R}^N$ since this case is explicitly used in Chap. 2. A sequence $\{f_m\}$ in $L^\infty(\Omega, L^q(U))$ $*$-weakly converges to $f \in L^\infty(\Omega, L^q(U))$ as $m \to \infty$ if and only if

$$\lim_{m \to \infty} \int_0^T \int_U f_m(x, t)\varphi(x, t)\mathrm{d}x\,\mathrm{d}t = \int_0^T \int_U f(x, t)\varphi(x, t)\mathrm{d}x\,\mathrm{d}t$$

for $\varphi \in L^1\left(\Omega, L^{q'}(U)\right)$. (Note that the space $L^p(\Omega, L^q(U))$ is identified with the space of all measurable functions $\varphi$ on $\Omega \times U$ such that $\int_0^T \|\varphi\|_{L^q(U)}^p(t)\,\mathrm{d}t < \infty$ or $\int_0^T \left(\int_U |\varphi(x, t)|^q\,\mathrm{d}x\right)^{p/q}\mathrm{d}t < \infty$ for $1 \le p, q < \infty$.)

(5) Besides the basic properties of the Lebesgue integrals, we frequently use a few estimates involving $L^p$-norms. These properties are by now standard and found in many books, including [19]. For example, we frequently use the *Hölder inequality*

$$\|fg\|_p \le \|f\|_r \|g\|_q$$

with $1/p = 1/r + 1/q$ for $f \in L^r(\Omega)$, $g \in L^q(\Omega)$, where $p, q, r \in [1, \infty]$. Here, we interpret $1/\infty = 0$. In the case $p = 1$, $r = q = 2$, this inequality is called the *Schwarz inequality*. As an application, we have *Young's inequality* for a convolution

$$\|f * g\|_p \le \|f\|_q \|g\|_r$$

for $f \in L^q(\mathbf{R}^N)$, $g \in L^r(\mathbf{R}^N)$ with $1/p = 1/q + 1/r - 1$ and $p, q, r \in [1, \infty]$; see, for example, [45, Chapter 4]. In this book, we use this inequality when $\mathbf{R}^N$ is replaced by $\mathbf{T}^N$.

(6) In analysis, we often need an approximation of a function by smooth functions. We only recall an elementary fact. The space $C_c^\infty(\Omega)$ is dense in $L^p(\Omega)$ for $p \in [1, \infty)$; see, for example, [19, Corollary 4.23]. However, it is not dense in $L^\infty(\Omega)$.

# Bibliography

1. R.P. Agarwal, V. Lakshmikantham, *Uniqueness and Nonuniqueness Criteria for Ordinary Differential Equations*. Series in Real Analysis, vol. 6 (World Scientific Publishing, River Edge, NJ, 1993)
2. L. Ambrosio, Transport equation and Cauchy problem for $BV$ vector fields. Invent. Math. **158**, 227–260 (2004)
3. L. Ambrosio, N. Gigli, G. Savaré, *Gradient Flows in Metric Spaces and in the Space of Probability Measures*, 2nd edn. Lectures in Mathematics ETH Zürich. (Birkhäuser Verlag, Basel, 2008)
4. L. Ambrosio, M. Lecumberry, S. Maniglia, Lipschitz regularity and approximate differentiability of the DiPerna-Lions flow. Rend. Sem. Mat. Univ. Padova **114**, 29–50 (2005)
5. F. Andreu-Vaillo, V. Caselles, J. M. Mazón, *Parabolic Quasilinear Equations Minimizing Linear Growth Functionals*. Progress in Mathematics, vol. 223 (Birkhäuser Verlag, Basel, 2004)
6. V. Barbu, *Nonlinear Differential Equations of Monotone Types in Banach Spaces*. Springer Monographs in Mathematics (Springer, New York, 2010), x+272 pp
7. M. Bardi, I. Capuzzo-Dolcetta, *Optimal Control and Viscosity Solutions of Hamilton-Jacobi-Bellman Equations*. With appendices by Maurizio Falcone and Pierpaolo Soravia. Systems & Control: Foundations & Applications (Birkhäuser Boston, Boston, MA, 1997)
8. M. Bardi, M.G. Crandall, L.C. Evans, H.M. Soner, P.E. Souganidis, *Viscosity Solutions and Applications*. Lecture Notes in Mathematics, vol. 1660 (Springer, Berlin, 1997)
9. G. Barles, *Solutions de viscosité des équations de Hamilton-Jacobi*. Mathématiques & Applications (Berlin), vol. 17 (Springer, Paris, 1994)
10. G. Barles, H.M. Soner, P.E. Souganidis, Front propagation and phase field theory. SIAM J. Control Optim. **31**, 439–469 (1993)
11. E.N. Barron, R. Jensen, Semicontinuous viscosity solutions for Hamilton-Jacobi equations with convex Hamiltonians. Commun. Partial Differ. Equ. **15**, 1713–1742 (1990)
12. E.N. Barron, R. Jensen, Optimal control and semicontinuous viscosity solutions. Proc. Am. Math. Soc. **113**, 397–402 (1991)
13. D. Bothe, On moving hypersurfaces and the discontinuous ODE-system associated with two-phase flows. Nonlinearity **33**(10), 5425–5456 (2020)
14. Y. Brenier, Averaged multivalued solutions for scalar conservation laws. SIAM J. Numer. Anal. **21**, 1013–1037 (1984)
15. Y. Brenier, $L^2$ formulation of multidimensional scalar conservation laws. Arch. Ration. Mech. Anal. **193**, 1–19 (2009)
16. A. Bressan, The unique limit of the Glimm scheme. Arch. Ration. Mech. Anal. **130**, 205–230 (1995)
17. A. Bressan, *Hyperbolic Systems of Conservation Laws. The One-Dimensional Cauchy Problem*. Oxford Lecture Series in Mathematics and Its Applications, vol. 20 (Oxford University Press, Oxford, 2000)

© The Author(s), under exclusive license to Springer Nature Switzerland AG 2023      149
M.-H. Giga, Y. Giga, *A Basic Guide to Uniqueness Problems for Evolutionary Differential Equations*, Compact Textbooks in Mathematics,
https://doi.org/10.1007/978-3-031-34796-2

18. H. Brézis, *Opérateurs maximaux monotones et semi-groupes de contractions dans les espaces de Hilbert*. North-Holland Mathematics Studies, No. 5. Notas de Matemática (50) (North-Holland Publishing, Amsterdam-London; American Elsevier Publishing, New York, 1973)

19. H. Brézis, *Functional Analysis, Sobolev Spaces and Partial Differential Equations*. Universitext (Springer, New York, 2011)

20. T. Buckmaster, V. Vicol, Nonuniqueness of weak solutions to the Navier-Stokes equation. Ann. Math. (2) **189**, 101–144 (2019)

21. Y.G. Chen, Y. Giga, S. Goto, Uniqueness and existence of viscosity solutions of generalized mean curvature flow equations. J. Differ. Geom. **33**, 749–786 (1991); Announcement Proc. Jpn. Acad. Ser. A Math. Sci. **65**, 207–210 (1989)

22. Y.G. Chen, Y. Giga, S. Goto, Remarks on viscosity solutions for evolution equations. Proc. Jpn. Acad. Ser. A Math. Sci. **67**, 323–328 (1991)

23. J.Á. Cid, R.L. Pouso, A generalization of Montel-Tonelli's uniqueness theorem. J. Math. Anal. Appl. **429**, 1173–1177 (2015)

24. A. Constantin, On Nagumo's theorem. Proc. Jpn. Acad. Ser. A Math. Sci. **86**, 41–44 (2010)

25. A. Constantin, A uniqueness criterion for ordinary differential equations. J. Diff. Equ. **342**, 179–192 (2023)

26. M.G. Crandall, H. Ishii, P.-L. Lions, User's guide to viscosity solutions of second order partial differential equations. Bull. Am. Math. Soc. (N.S.) **27**, 1–67 (1992)

27. M.G. Crandall, T.M. Liggett, A theorem and a counterexample in the theory of semigroups of nonlinear transformations. Trans. Am. Math. Soc. **160**, 263–278 (1971)

28. M.G. Crandall, P.-L. Lions, Condition d'unicité pour les solutions généralisées des équations de Hamilton-Jacobi du premier ordre. C. R. Acad. Sci. Paris Sér. I Math. **292**, 183–186 (1981)

29. M.G. Crandall, P.-L. Lions, Viscosity solutions of Hamilton-Jacobi equations. Trans. Am. Math. Soc. **277**, 1–42 (1983)

30. G. Crippa, C. De Lellis, Estimates and regularity results for the DiPerna-Lions flow. J. Reine Angew. Math. **616**, 15–46 (2008)

31. K. Deimling, *Multivalued Differential Equations*. De Gruyter Series in Nonlinear Analysis and Applications, vol. 1 (Walter de Gruyter & Co., Berlin, 1992)

32. R.J. DiPerna, P.-L. Lions, Ordinary differential equations, transport theory and Sobolev spaces. Invent. Math. **98**, 511–547 (1989)

33. S.S. Dragomir, *Some Gronwall Type Inequalities and Applications* (Nova Science Publishers, Hauppauge, NY, 2003)

34. T.D. Drivas, T.M. Elgindi, G. Iyer, I.-J. Jeong, Anomalous dissipation in passive scalar transport. Arch. Ration. Mech. Anal. **243**, 1151–1180 (2022)

35. N. Dunford, J.T. Schwartz, *Linear Operators. Part I. General Theory*. Wiley Classics Library. A Wiley-Interscience Publication (Wiley, New York, 1988), xiv+858 pp

36. L.C. Evans, *Partial Differential Equations*, 2nd edn. Graduate Studies in Mathematics, vol. 19. (American Mathematical Society, Providence, RI, 2010)

37. L.C. Evans, J. Spruck, Motion of level sets by mean curvature. I. J. Differ. Geom. **33**, 635–681 (1991)

38. A. Fathi, A. Siconolfi, PDE aspects of Aubry-Mather theory for quasiconvex Hamiltonians. Calc. Var. Partial Differ. Equ. **22**, 185–228 (2005)

39. E. Feireisl, A. Novotný, *Singular Limits in Thermodynamics of Viscous Fluids*, 2nd edn. Advances in Mathematical Fluid Mechanics (Birkhäuser/Springer, Cham, 2017)

40. A.F. Filippov, *Differential Equations with Discontinuous Right-Hand Sides, and Differential Inclusions*. Nonlinear Analysis and Nonlinear Differential Equations (FizMatLit, Moscow, 2003), pp. 265–288

41. W.H. Fleming, H.M. Soner, *Controlled Markov Processes and Viscosity Solutions*, 2nd edn. Stochastic Modelling and Applied Probability, vol. 25 (Springer, New York, 2006)

42. G. B. Folland, *Real Analysis*. Modern Techniques and Their Applications. Pure and Applied Mathematics (New York). A Wiley-Interscience Publication (Wiley, New York, 1984)

43. M.-H. Giga, Y. Giga, Evolving graphs by singular weighted curvature. Arch. Ration. Mech. Anal. **141**, 117–198 (1998)

44. M.-H. Giga, Y. Giga, *Minimal Vertical Singular Diffusion Preventing Overturning for the Burgers Equation*. Recent Advances in Scientific Computing and Partial Differential Equations (Hong Kong, 2002). Contemp. Math., vol. 330 (Amer. Math. Soc., Providence, RI, 2003), pp. 73–88

45. M.-H. Giga, Y. Giga, J. Saal, *Nonlinear Partial Differential Equations. Asymptotic Behavior of Solutions and Self-Similar Solutions*. Progress in Nonlinear Differential Equations and Their Applications, vol. 79 (Birkhäuser Boston, Boston, MA, 2010)

46. Y. Giga, Viscosity solutions with shocks. Commun. Pure Appl. Math. **55**, 431–480 (2002)

47. Y. Giga, *Surface Evolution Equations. A Level Set Approach*. Monographs in Mathematics, vol. 99 (Birkhäuser Verlag, Basel, 2006)

48. Y. Giga, A. Kubo, H. Kuroda, J. Okamoto, K. Sakakibara, M. Uesaka, Fractional time differential equations as a singular limit of the Kobayashi-Warren-Carter system (2023). arXiv: 2306.15235

49. Y. Giga, T. Miyakawa, A kinetic construction of global solutions of first order quasilinear equations. Duke Math. J. **50**, 505–515 (1983)

50. Y. Giga, T. Miyakawa, S. Oharu, A kinetic approach to general first order quasilinear equations. Trans. Am. Math. Soc. **287**, 723–743 (1985)

51. Y. Giga, M.-H. Sato, A level set approach to semicontinuous viscosity solutions for Cauchy problems. Commun. Partial Differ. Equ. **26**, 813–839 (2001)

52. P. Hartman, *Ordinary Differential Equations*. Reprint of the second edition. (Birkhäuser, Boston, MA, 1982)

53. H. Holden, N.H. Risebro, *Front Tracking for Hyperbolic Conservation Laws*, 2nd edn. Applied Mathematical Sciences, vol. 152 (Springer, Heidelberg, 2015)

54. E. Hopf, The partial differential equation $u_t + uu_x = \mu u_{xx}$. Commun. Pure Appl. Math. **3**, 201–230 (1950)

55. L. Huysmans, E.S. Titi, Non-uniqueness and inadmissibility of the vanishing viscosity limit of the passive scalar transport equation (2023). arXiv: 2307.00809

56. H. Ishii, Hamilton-Jacobi equations with discontinuous Hamiltonians on arbitrary open sets. Bull. Fac. Sci. Eng. Chuo Univ. **28**, 33–77 (1985)

57. H. Ishii, Perron's method for Hamilton-Jacobi equations. Duke Math. J. **55**, 369–384 (1987)

58. H. Ishii, A simple, direct proof of uniqueness for solutions of the Hamilton-Jacobi equations of eikonal type. Proc. Am. Math. Soc. **100**, 247–251 (1987)

59. H. Ishii, Viscosity solutions of nonlinear partial differential equations [translation of Sūgaku **46**, 144–157 (1994)]. Sugaku Expositions **9**, 135–152 (1996)

60. H. Ishii, H. Mitake, Representation formulas for solutions of Hamilton-Jacobi equations with convex Hamiltonians. Indiana Univ. Math. J. **56**, 2159–2183 (2007)

61. Y. Kobayashi, N. Tanaka, Semigroups of Lipschitz operators. Adv. Differ. Equ. **6**, 613–640 (2001)

62. Y. Kobayashi, N. Tanaka, Y. Tomizawa, Nonautonomous differential equations and Lipschitz evolution operators in Banach spaces. Hiroshima Math. J. **45**, 267–307 (2015)

63. S. Koike, *A Beginner's Guide to the Theory of Viscosity Solutions*. MSJ Memoirs, vol. 13 (Mathematical Society of Japan, Tokyo, 2004)

64. S. Koike, *Nenseikai. (Japanese) [Viscosity Solutions]* Kyoritsu Kouza Sūgaku No Kagayaki, vol. 16. (Kyoritsu Shuppan Co., Tokyo, 2016)

65. Y. Kōmura, Nonlinear semi-groups in Hilbert space. J. Math. Soc. Jpn. **19**, 493–507 (1967)

66. Y. Kōmura, Y. Konishi, *Hisenkei hatten hoteishiki. (Japanese) [Nonlinear Evolution Equations]*, 2nd edn. Iwanami Shoten Kiso Sūgaku [Iwanami Lectures on Fundamental Mathematics], vol. 16. Kaisekigaku (II) [Analysis (II)], vii (Iwanami Shoten, Tokyo, 1983)

67. S.G. Krantz, H.R. Parks, *The Implicit Function Theorem. History, Theory, and Applications*. Reprint of the 2003 edition. Modern Birkhäuser Classics (Birkhäuser/Springer, New York, 2013)

68. S.N. Kružkov, First order quasilinear equations with several independent variables. (Russian) Mat. Sb. (N.S.) **81**(123), 228–255 (1970)

69. A. Kubica, K. Ryszewska, M. Yamamoto, Time-fractional differential equations – a theoretical introduction. Springer Briefs Math. (Springer, Singapore, 2020), x+134 pp.
70. O.A. Ladyženskaja, V.A. Solonnikov, N.N. Ural'ceva, Linear and quasi-linear equations of parabolic type. (American Mathematical Society, Providence, RI, 1968)
71. P.-L. Lions, *Generalized Solutions of Hamilton-Jacobi Equations.* Research Notes in Mathematics, vol. 69 (Pitman (Advanced Publishing Program), Boston, MA, London, 1982)
72. A. Lunardi, *Analytic Semigroups and Optimal Regularity in Parabolic Problems.* [2013 reprint of the 1995 original]. Modern Birkhäuser Classics (Birkhäuser/Springer Basel AG, Basel, 1995), xviii+424 pp
73. G. J. Minty, Monotone (nonlinear) operators in Hilbert space. Duke Math. J. **29**, 341–346 (1962)
74. I. Miyadera, *Hisenkei Hangun.* (Japanese), Kinokuniya, Tokyo, 1977. English translation: Nonlinear semigroups. Translations of Mathematical Monograph, vol. 109 (Translated by Choong Yun Cho) (American Mathematical Society, Providence, RI, 1992), viii+231 pp.
75. S. Modena, L. Székelyhidi, Jr., Non-uniqueness for the transport equation with Sobolev vector fields. Ann. PDE **4**, Paper No. 18, 38 pp. (2018)
76. H. Morimoto, *Stochastic Control and Mathematical Modeling.* Applications in Economics. Encyclopedia of Mathematics and Its Applications, vol. 131 (Cambridge University Press, Cambridge, 2010)
77. M. Nagumo, Eine hinreichende Bedingung für die Unität der Lösung von Differentialgleichungen erster Ordnung. Jpn. J. Math. Trans. Abstracts **3**, 107–112 (1926)
78. H. Okamoto, Navier-Stokes Houteishiki No Sūri. (Japanese) [Mathematical analysis of the Navier-Stokes equations] (University of Tokyo Press, Tokyo, 2009)
79. H. Okamura, Condition nécessaire et suffisante remplie par les équations différentielles ordinaires sans points de Peano. Mem. Coll. Sci. Kyoto Imp. Univ. Ser. A. **24**, 21–28 (1942)
80. W.F. Osgood, Beweis der Existenz einer Lösung der Differentialgleichung $\frac{dy}{dx} = f(x, y)$ ohne Hinzunahme der Cauchy-Lipschitz'schen Bedingung. Monatsh. Math. Phys. **9**, 331–345 (1898)
81. S. Osher, J.A. Sethian, Fronts propagating with curvature-dependent speed: algorithms based on Hamilton-Jacobi formulations. J. Comput. Phys. **79**, 12–49 (1988)
82. M. Palani, C.C. Tisdell, A. Usachev, Qualitative results for solutions to nonlinear Caputo differential equations satisfying the Osgood condition. Fract. Differ. Calc. **8**, 151–164 (2018)
83. I. Podlubny, *Fractional Differential Equations. An Introduction to Fractional Derivatives, Fractional Differential Equations, to Methods of Their Solution and Some of Their Applications.* Mathematics in Science and Engineering, vol. 198 (Academic Press, San Diego, CA, 1999)
84. M.H. Protter, H.F. Weinberger, *Maximum Principles in Differential Equations.* Corrected reprint of the 1967 original (Springer, New York, 1984), x+261 pp.
85. L. Tonelli, Sull'unicità della soluzione di un'equazione differenziale ordinaria. Rend. Acc. Naz. Lincei **6**, 272–277 (1925)
86. T. Tsuruhashi, T. Yoneda, Microscopic expression of anomalous dissipation in passive scalar transport (2022). arXiv: 2212.06395
87. H.V. Tran, *Hamilton-Jacobi Equations – Theory and Applications.* Graduate Studies in Mathematics, vol. 213. (American Mathematical Society, Providence, RI, 2021)
88. Y.-H.R. Tsai, Y. Giga, S. Osher, A level set approach for computing discontinuous solutions of Hamilton-Jacobi equations. Math. Comput. **72**, 159–181 (2003)
89. W. Wolibner, Un theorème sur l'existence du mouvement plan d'un fluide parfait, homogène, incompressible, pendant un temps infiniment long. Math. Z. **37**, 698–726 (1933)
90. K. Yosida, *Functional Analysis.* Reprint of the sixth (1980) edition. Classics in Mathematics. (Springer, Berlin, 1995)

# Index

© The Author(s), under exclusive license to Springer Nature Switzerland AG 2023
M.-H. Giga, Y. Giga, *A Basic Guide to Uniqueness Problems for Evolutionary Differential Equations*, Compact Textbooks in Mathematics,
https://doi.org/10.1007/978-3-031-34796-2